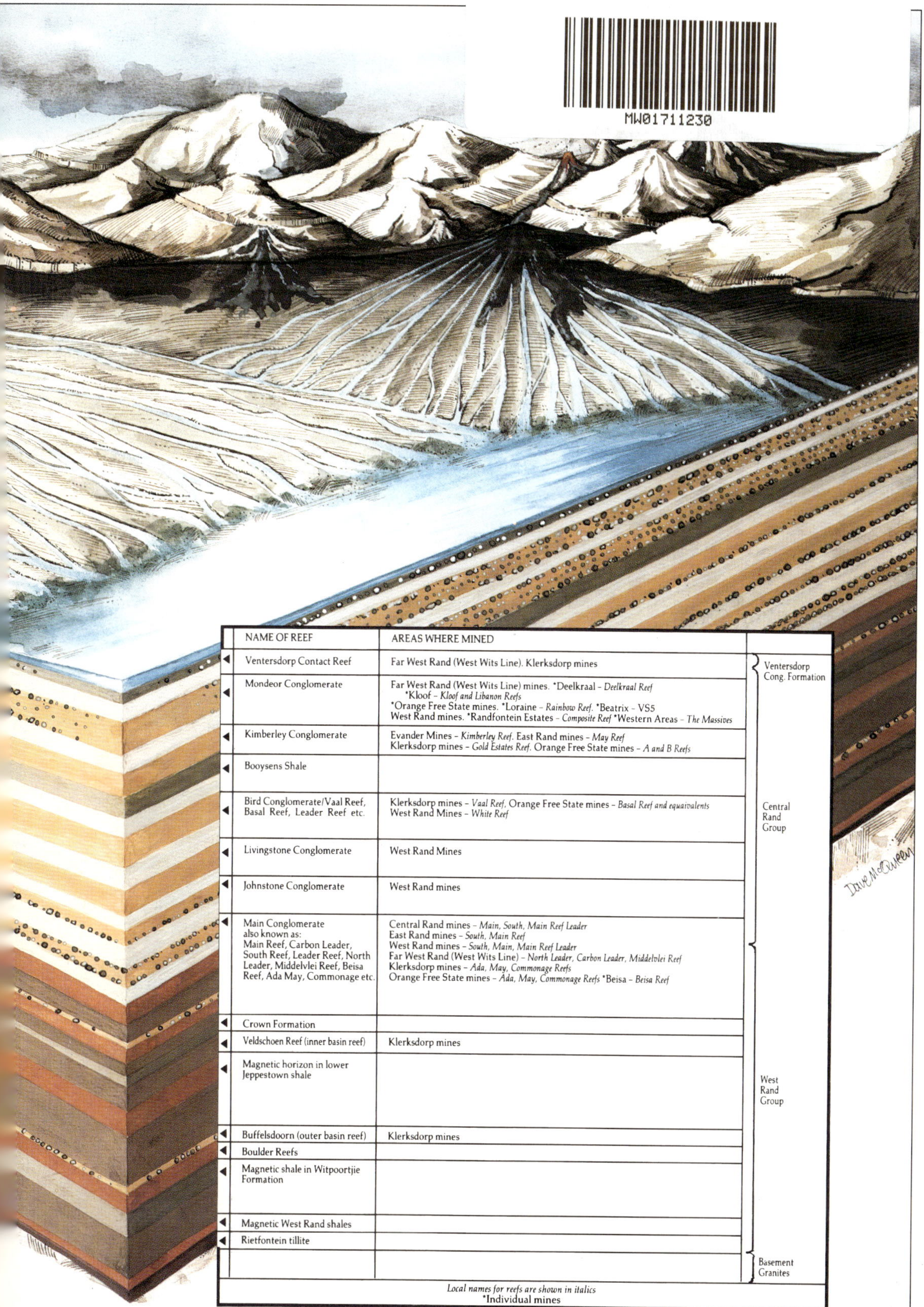

	NAME OF REEF	AREAS WHERE MINED	
◀	Ventersdorp Contact Reef	Far West Rand (West Wits Line). Klerksdorp mines	Ventersdorp Cong. Formation
◀	Mondeor Conglomerate	Far West Rand (West Wits Line) mines. *Deelkraal – *Deelkraal Reef* *Kloof – *Kloof and Libanon Reefs* *Orange Free State mines. *Loraine – *Rainbow Reef*. *Beatrix – VS5 West Rand mines. *Randfontein Estates – *Composite Reef* *Western Areas – *The Massives*	
◀	Kimberley Conglomerate	Evander Mines – *Kimberley Reef*. East Rand mines – *May Reef* Klerksdorp mines – *Gold Estates Reef*. Orange Free State mines – *A and B Reefs*	
◀	Booysens Shale		
◀	Bird Conglomerate/Vaal Reef, Basal Reef, Leader Reef etc.	Klerksdorp mines – *Vaal Reef*, Orange Free State mines – *Basal Reef and equaivalents* West Rand Mines – *White Reef*	Central Rand Group
◀	Livingstone Conglomerate	West Rand Mines	
◀	Johnstone Conglomerate	West Rand mines	
◀	Main Conglomerate also known as: Main Reef, Carbon Leader, South Reef, Leader Reef, North Leader, Middelvlei Reef, Beisa Reef, Ada May, Commonage etc.	Central Rand mines – *Main, South, Main Reef Leader* East Rand mines – *South, Main Reef* West Rand mines – *South, Main, Main Reef Leader* Far West Rand (West Wits Line) – *North Leader, Carbon Leader, Middelvlei Reef* Klerksdorp mines – *Ada, May, Commonage Reefs* Orange Free State mines – *Ada, May, Commonage Reefs* *Beisa – *Beisa Reef*	
◀	Crown Formation		
◀	Veldschoen Reef (inner basin reef)	Klerksdorp mines	
◀	Magnetic horizon in lower Jeppestown shale		West Rand Group
◀	Buffelsdoorn (outer basin reef)	Klerksdorp mines	
◀	Boulder Reefs		
◀	Magnetic shale in Witpoortjie Formation		
◀	Magnetic West Rand shales		
◀	Rietfontein tillite		
			Basement Granites

Local names for reefs are shown in italics
*Individual mines

DOWN
WHERE NO LION WALKED
THE STORY OF WESTERN DEEP LEVELS

DOWN WHERE NO LION WALKED

THE STORY OF WESTERN DEEP LEVELS

JOHN OXLEY

SOUTHERN
BOOK PUBLISHERS

ISBN 1 86812 202 6

First edition, first impression 1989

Published by
Southern Book Publishers (Pty) Ltd
PO Box 548, Bergvlei 2012
Johannesburg

Cover design by William Steyn

Set in 11 on 13 pt Andover
Reproduction
by Unifoto, Cape Town
Printed and bound by CTP Book Printers, Cape

BD9752

A Hymn in Praise of Wisdom

Silver has its mines,
and gold a place for refining.
Iron is extracted from the earth,
the smelted rocks yield copper.
Man makes an end to darkness
when he pierces the uttermost depths
the black and lightless rock.
Mines the lamp-folk dig
in places where there is no foothold,
and hang suspended far from mankind.
That earth from which bread comes
is ravaged underground by fire.
Down there the rocks are set with sapphires,
full of spangles of gold.
Down there is a path unknown to birds of prey,
unseen by the eye of any vulture;
a path not trodden by the lordly beasts,
where no lion ever walked.
Man attacks its flinty sides,
upturning mountains by their roots,
driving tunnels through the rocks,
on the watch for anything precious.
He explores the sources of rivers,
and brings to daylight secrets that were hidden.
But tell me, where does wisdom come from ?
Where is understanding to be found ?

JOB: Chapter 28 verses 1–12

(Modern version: Jerusalem Bible)

To
the men and their families who made
Western Deep Levels
great.

Contents

WESTERN DEEP LEVELS
Chairmen
1957–1987

Sir Ernest Oppenheimer
1957

H.F. Oppenheimer
1958–1963

H.C. Koch
1963–1969

J.W. Shilling
1969–1974

D.A. Etheredge, O.B.E.
1974–1977

G. Langton
1978–1983

E.P. Gush
1983–

WESTERN DEEP LEVELS

Managing Directors
1957–1987

John Shilling
1957–1962

Hermann Koch
1963

Douglas Hoffe
1963–1964

Rawdon Lawrie
1974–1985

Theo Pretorius
1985–

PHOTO CREDITS
Anglo American Corporation photographic department; Black and white:
pages: x, xi, xii, xiii, xvi, 17, 70, 83, 85, 95, 98, 104, 110, 112, 156, 160, 172, 185, 190, 200, 211, 224,
All colour pictures except first one.

Chamber of Mines;
Endpapers.
Black and white: xi.

SPIX International;
First colour picture.
Black and white: 7, 171.

Drawings by Professor Jack Grossert, who also designed and drew the decorative initials with which each chapter commences.

LIST OF ILLUSTRATIONS

Endpapers

Colour section in chronological order

Black and white illustrations

24 July 1957. Western Transvaal: Sir Ernest Oppenheimer, Chairman of the Anglo American Corporation of South Africa, sets in motion the pre-grouting drills, marking the inauguration of the Western Deep Levels mine. On right: Frank Ludwig, first Manager of the mine.

Foreword

H. F. Oppenheimer

I am very pleased that the history of Western Deep Levels should have been written, and written in such an interesting way. John Oxley has placed this record of technical innovation and skill, financial ingenuity and daring, against a wide and fascinating background, scientific, historical and human. Western Deeps has a very special interest for me, both for business and personal reasons. It is the largest and boldest single mining project ever undertaken by the Anglo American group and it was the last major economic development in which my father was able to take a direct part. He became the first Chairman of the company and looked forward to the mine's being brought to production with enormous pleasure and with a degree of confidence in the outcome which was not always shared by some of the younger members of our team.

In the Anglo American group we have always looked on Western Deeps with particular pride as one of our outstanding achievements. It would not be right, however, to celebrate any major success in the development of the "West Wits line" without paying tribute to the courage, skill and imagination of the Consolidated Goldfields group (now Goldfields of South Africa) who were responsible in the first place for the opening up of this great new goldfield. When it came to the Western Deeps project, there was healthy competition between the Anglo American and Goldfields groups, but in the end there was also a degree of co-operation between them and with the Rand Mines group, which was important for the success of the project. It was an example of the way in which the so-called "group system" works to the advantage of the mining industry as a whole and of the country.

Mining at depths greater than in the past brought fresh technical and financial problems, but it had long been obvious of course that, if those problems were solved, there existed down-dip from the established mines an open frontier for the expansion of the gold mining industry. Mining at such great depths would necessarily involve unusually high capital expenditure and working costs and therefore could only be economically undertaken if the ore in the area could be expected to be of exceptionally high grade. Dr J. Austen Bancroft, the Anglo American's Consulting Geologist and a man of remarkable character and ability, identified the area to the south of the West Driefontein and Blyvooruitzicht mines as being the most promising available, particularly since it had the added attraction that "Ventersdorp Contact Reef" also occurred in the area at a shallower depth than the "Carbon Leader" on which reef the new mine would mainly depend. Western Deeps could therefore be planned as a double-decker mine and this was an essential factor for the economic viability of the venture. No doubt it had generally been expected that at some distant date ore down-dip from the southern boundaries of West Driefontein and Blyvooruitzicht would be mined, but it had been supposed that this could only take place by gradual extensions from existing workings. The originality of the initiative taken by Anglo American lay in the perception that this area could be tackled as an entirely new proposition, with mining of the "Carbon Leader" starting at depths far greater than had ever previously been contemplated as the basis for the establishment of a new mine. Anglo American proceeded by forming a prospecting company with the bold title of "Western Ultra Deep Levels" to take up and explore land to the south of the farms on which the existing mines had been established. The expectation was that, if drilling results were favourable, it should be possible to obtain a mining lease to exploit not only ground held by "Western Ultra Deeps" but also the southern parts of the farms Driefontein and Blyvooruitzicht which, because of the depth of the reef, had not been included in the West Driefontein and Blyvooruitzicht mines. And so matters turned out.

But that was not all. The great delays in getting to the reefs at those depths involved the investment of exceptionally large amounts of capital for a long period before any amortisation or profits were earned. From the government we sought an allowance on invested capital to recompense the mine owners — and shareholders — for this unduly long delay in financial returns.

It is greatly to the credit of the government that they quickly saw the value and the wisdom of such a concession. This was the initiation of the present well-known capital allowance system in the structure of the country's present gold mining taxation.

It has been said that fortune follows the brave and also that fortune favours fools. Whether it was for one or the other of these reasons I cannot tell, but anyhow a great and successful mine was established and the frontier pushed forward into new lands for the whole mining industry.

I congratulate John Oxley on his book. It tells with clarity and elegance the story of a great technical and business adventure and succeeds too in catching that spirit of romance which is an essential part of the mining industry.

Preface

This is the story of a gold-mine — a great gold-mine — and of the region where it is situated, as well as about the tens of thousands of men of various races who, through its short life to date, have laboured to make it great. Without the dedication and the contributions of them all, each in his fashion, fame and success could not have come its way, nor achievement so richly gilded its record.

It is a splendid tale to tell: the birth and development of an organism which, predestined, had its genesis in that patch of Africa's earth hundreds of millions of years ago when the world itself was young, aeons before life began and man's slow evolution commenced.

The mine is Western Deep Levels, situated on the Transvaal highveld in South Africa about 70 kilometres west of Johannesburg, a city itself made famous by gold, a city which sits astride a rich reef that, since the precious metal was first discovered there in 1886, has yielded more than three-quarters of all the gold man has mined during his millennia upon earth.

The story could conceivably be told in several ways. For example, the progress of the mine and its achievements could perhaps be set out in a series of diagrams and statistics — mere drawings and numerical tables: and an impressive saga those tidy graphics would indeed present for the scientist and technician, the engineer and the economist.

But for many such a presentation would be a stark and sterile one, devoid of the essential, vibrant human content. Furthermore, such an account would be unforgivably silent on the immensity of the role which men have played in the creation of a rich industrial complex, whose endeavours have established those very records and statistics. Nor would mere graphs and figures tell of the struggles and the triumphs — of the human aches and the pains — which brought the victories, hard won along the way.

No less intriguing is the story, controversial though it may be, of how nature had created this rich gold deposit — mines waiting to be discovered! In this sense, Western Deep Levels may be regarded as the product of history. Its ultimate creation was pre-ordained by the laying down of the gold deposits millions of years ago and its development was predestined from the moment nomadic man put a value upon gold. Man's ingenuity developed to exploit the value. The presence there and the current activity of Western Deep Levels can perhaps thus be said now to be merely fulfilling a long-prescribed role — in the time-slot of the twentieth and twenty-first centuries — in the continuum of an infinite history. The present is conceived in the womb of the past: todays are fashioned by their yesterdays.

In geological and archaeological terms, the unique and varied richness of the earth in which Western Deep Levels is situated is too fascinating to be ignored; while, in more contemporary terms, for the past century and a half the surrounding countryside has also seen much that is significant in the development of modern South Africa.

In a sense, every book mirrors something of the particular interests of its author, revealing partiality and thus perhaps laying him open to charges of bias. The risk of such accusations is heightened when the options for such predilection are widened by the scope of the subject — as in this study which ranges over an immensely wide spectrum of different disciplines, each, though connected to a central theme like spokes to the hub of a wheel, so different from the rest.

Writing this account may perhaps be likened to mining itself: for just as thousands of tons of rock have to be processed to produce a bar of gold, so considerable research was necessary in collecting material for this book. Specks of gold lie unseen in the remotest rock.

A study of Western Deep Levels mine — its conception, its planning, its establishment and development — requires, I found, more than a rudimentary appreciation and knowledge of finance, economics and company law, as well as of geology, of engineering and metallurgy, besides, additionally, some essential comprehension of a host of other technical subjects, mostly esoteric for the layman and easily comprehended only by their particular devotees and attendant specialists.

Into these personally unknown and uncharted territories I ventured as a timid trespasser.

"... Much have I travelled in the realms of gold,
And many goodly states and kingdoms seen; ..."

In each region I was fortunate enough to find the natives friendly; and in each of the separate imperial domains, with the patient help of their masters and their mandarins, I was enabled to garner much of the material required for this story.

Not all problems ceased there, however. It has been necessary to weave that material into a rounded story which I have tried to make intelligible to the layman and of interest to the specialist and scientist. At the same time, furthermore, in each instance I have had to steer a course which, I hope, avoids favouring any one discipline at the expense of others; and, indeed, ensures that any single facet — historical or contemporary — has not been more brightly polished than the rest.

When Western Deep Levels was established, it was calculated that the mine would have a working life of some sixty years. This account has been compiled thirty years after Sir Ernest Oppenheimer's symbolic launching of operations in July 1957. From this half-way position, then, one looks back and also forward, at least to an equal span of years ahead.

Not only has Western Deeps already written itself into the Guinness Book of Records by sinking the deepest mine shaft in history, but in its short life it has also achieved an impressive number of other records, as this story relates. It is technologically the most advanced mine in the world, and one of the richest. Already it has fulfilled Sir Ernest's prediction that Western Deeps was destined to be "the most important enterprise ever started in South Africa." Opportunities for yet further achievements beckon brightly across the years ahead.

Western Deep Levels is an impressive memorial to the man who, above all others, conceived it: Sir Ernest Oppenheimer. It remains a monument to the men whose dedication, ingenuity, industry and courage created it.

The knowledge of the various experts consulted is too profound to be fallible. To them I am grateful. For whatever errors and misrepresentations this account may unwittingly contain, however, I alone must be deemed responsible, as I am also for any significant omissions.

J.O.

Pretoria, February 1988

CHAPTER ONE

The deepest Mine in the World

Western Deep Levels gold-mine, situated some 70 kilometres west of Johannesburg on the South African highveld, is one of more than forty mines which, following the course of the rich auriferous reef, punctuate a broad, irregular and undulating semi-circle that sweeps almost 500 kilometres along the Witwatersrand and the western Transvaal and reaches southward into the wide and extensive plains of the northern Orange Free State.

This is the region, richly endowed by nature, which, since the discovery of gold there a little more than a century ago, has produced most of the gold man has known in all his millennia upon earth.

The fame of Western Deep Levels rests not upon its age nor upon its size, but because it is the deepest mine in the world, an achievement acknowledged in the Guinness Book of Records where the mine's Number 2 Shaft is described as "the world's deepest mine at 3 582 metres, at which depth the ambient rock temperature is recorded as 131 degrees F or 55 degrees C."

To keep pace with continuing developments at Western Deep Levels, future editions will surely have to amend the entry again — and yet again — on and on, as, during the lifetime of the mine, man's appetite for gold continues and grows. A new shaft complex, initiated in the 1980s, is already on its way to even greater depths, making fresh history as the mine sets out to challenge the future.

That somewhat bald, laconic statement in the record book tells merely of the depth achieved. It is left to the imagination to contemplate the immense effort which created the record, to consider the brains which shaped the concept and planned the project and to imagine the heart, the muscle and the physical exertion which brought it to fruition.

This record was not achieved haphazardly. It was not that, in reaching down for the gold, the mine happened to attain such depths. The levels at which mining would take place on Western Deeps were carefully investigated and operations were planned before the site was cleared, before drills began twisting their way downwards and, in search of the gold known to be there, gouging into the hard rock underground.

The manifold risks and dangers were carefully weighed.

But while for the man in the street the record depth itself excites the mind, for miners and scientists everywhere it is the manner of achievement which sets the mind boggling, and it is the conquering of the obstacles, large and small, along the way which excites their admiration.

In a historical perspective of their own, developments at Western Deeps have to be seen as having two phases. The first embraces the establishment of Shafts Numbers 2 and 3, with their manifold triumphs and records. The second phase commenced more recently — almost a quarter of a century after the first began — with the creation of a new — Number 1 — shaft. While in no way overlooking the records and achievements of the earlier shafts, it is now this new project, heir to the deep-level technology pioneered at the older shafts, on which the eyes of the mining world are focused.

The achievements of the earlier stage linked with plans for the succeeding phase bring the past and the future together in the present, providing a convenient point for a preliminary look at Western Deeps as a whole.

From the outset in the 1940s scientists and technicians had told of the unprecedented depths to which mining would have to go to reach the deep-lying gold-bearing reefs which had then been located in the Transvaal. The planners were not daunted, however, by the prospect of having to delve down to levels hitherto not worked by man. With faith in the techniques of the age and a belief in improving technology in the future, they set about creating a mine deeper than any before it, facing problems with ingenuity and with confidence born of ability.

For several decades the Anglo American Corporation of South Africa has been involved in a broad spectrum of mining — not only of gold, but also in coal, copper and other metals too, and was also not without experience in aspects of the mining of diamonds. Furthermore, that experience and those interests were spread over a number of countries. All this led to an accumulation of expertise of considerable breadth as well as depth within its ranks. It was this expertise, courageously utilised and exploited, which turned a vision into reality in creating and developing Western Deep Levels.

In achieving those record depths, new skills and innovative techniques were necessary to meet recurring fresh demands. Western Deeps pioneered many of them and its contribution to mining knowledge has been valuable to mining everywhere. Its initial two shafts have been laboratories where methods have been tested. The new Number 1 Shaft, utilising skills already developed and creating new techniques to meet fresh problems, is thus a bridge between the past and the future.

But however commendable the proud achievements of Western Deeps may be, these operations do relatively little more than prick the earth's crust — an outer layer which varies from between eight and 40 kilometres in thickness, encompassing a thick mantle which extends about 3 000 kilometres inward and, in turn, containing a molten stratum of a further 2 200 kilometres which itself surrounds a solid central core about 2 600 kilometres in diameter.

From his earliest beginnings man seems to have been concerned more with the sky above him than with the earth beneath. Until very recent times he has looked with greater interest and curiosity at the heavens, intrigued more by the wheeling stars in their courses and the silent planets of the night than by the possible contents of the earth on which his feet have rested. As a science astronomy outdates geology by many millennia.

The heavens provide man with light and with air to breathe and bless him with rains and with changing seasons for his sustenance. In the movements of the sun and the moon and the stars he has seen the ingredients of magic and from the mysteries of those planets and in the wonders of the seasons his worship originated. The earth, on the other hand, was hard and unyielding, dark and black.

For ancient man the invisibility of what lurked in the underground darkness implied danger, and in primitive minds the depths of the earth became associated with evil. For him at the earth's centre was Hades, the abode of the dead, the antithesis of the light-giving heaven. And yet further below Hades, the ancients Greeks believed, lay Tartarus, "a gloomy place in the underworld which lies as far distant from the earth as the earth does from the sky: it would take a falling anvil nine days to reach its bottom ..." It was where the wicked were despatched for punishment.

True, man did begin to scratch the surface of the earth by tilling, growing crops to sustain himself, while from time to time he made brave attempts to dig for whatever minerals he came to recognise from the samples which nature happened to provide on the surface. But even if man's hunger for these minerals tempted him to delve more deeply, he was thwarted by the inadequacy of his simple tools.

His early efforts to soar into the heavens would seem to have been more successful, however, and, according to legend, certainly to have demonstrated greater initiative.

If the myths are indeed accepted for their metaphorical truth, it was Daedalus, the Athenian, and Icarus, his son, who first broke contact with the earth and, in what the media today might describe as one of history's most spectacular gaol-breaks, reached into the air on man-made wings to escape from political captivity in Crete: but with unhappy, and indeed fatal, consequences.

Since then, though man has soared into space with greater sophistication and success, he has nevertheless been tardy in probing more deeply into the earth on which he lives.

Even in this twentieth century, with its established and developing technologies, man, still looking upwards, has travelled to the moon, some 400 000 kilometres into atmospheric space, passing through zones of heat and lethal radiation, through regions devoid of life-sustaining air, yet up to now he has delved less than four kilometres into the outer mantle of his own world, so much closer at hand. This has been done at Western Deep Levels mine. And while in recent years science has steered sensors almost eight-billion kilometres into the atmosphere, man's science has probed — by drilling — still less than 15 kilometres below the earth's surface: and that too only in recent decades.

Ironically perhaps, gold from the hard, dark earth is a vital component in the instruments used in the exploration of outer space.

But increasing sophistication of technology, together with a greater knowledge of the composition of the earth, now enables man to dig more deeply as he finds practical reasons — oil and minerals — which first tempt and then reward such exploration, and which, if the economic circumstances are right, urge advances to match the developing needs. The rewards are real and positive, and practical challenges breed progress.

It is this advancing technology, allied with human determination and courage, which has made possible the achievements at Western Deep Levels — the sinking of the world's deepest working shafts — in search of the gold which industry and hungry money markets everywhere demand.

At Western Deeps there are plans to go yet deeper: in the next few years to 4 000 metres, then perhaps in further stages to 5 000 and possibly even 6 000 metres below ground level where deposits of gold are known richly to lie: one day perhaps onward to 8 000 metres if favourable economic circumstances persist and developing technology can conquer the physical problems. Man's hopes match the technical promises of the future.

Just as Icarus — despite paternal warnings about flying too high —found his wings melting in the heat of the sun, so too is the thermal factor a major obstacle to underground exploration. Temperatures rise rapidly as excavations extend further below the surface. And even if advancing technology does meet — and overcome — this physical challenge, the factors of cost and related profits may prove to be the final hurdle in any commercial deep-mining enterprise.

Those problems belong to the future, however. Scientists are aware of what lies ahead, and at Western Deeps research planners are seeking solutions. Thirty pioneering years have bred a quiet confidence with which to tackle whatever problems may lie ahead.

Western Deep Levels Gold Mining Company Limited was incorporated in August 1957. It came into being after several years of planning and patient negotiations and, as part of those negotiations, after a mining lease had been granted in May of that year. Though of a fairly recent vintage and one of a younger generation of gold-mines, it is now one of Anglo American's major gold-mining projects.

The beginning of its operations were symbolised when, in July 1957, at the site of Number 2 Shaft, Sir Ernest Oppenheimer, Chairman of the Anglo American Corporation, pressed a button which set the first drills in motion. It was a simple ceremony which brought to reality, after fifteen years of complex planning, an enterprise that embodied Sir Ernest's vision and acumen and demonstrated his faith both in the gold industry and in South Africa's future.

Sadly, his death a few months later made his inauguration of the mine one of Sir Ernest's last public acts. The success of the Western Deeps gold-mine is thus a particular tribute and a fitting monument to this most distinguished South African.

From its beginning, Western Deep Levels developed rapidly. A long road lay ahead, and clearly there would be many barren years before gold was produced. The two initial shafts — Numbers 2 and 3 — developed to production in record time, however, and ahead of schedule. In less than five years, the mine had reached the shallower of the two main gold-bearing reefs in the area and was producing gold. Profits enabled the company to repay borrowed capital and provided financial resources for further developments — the exploitation of the richer Carbon Leader reef deeper down.

Success in the two decades which followed saw the consolidation of the company's financial position and, aided by a surge in the international price of gold, provided resources for further developments in the form of Number 1 Shaft, which has captured the imagination of the mining world. Its renown rests on its imaginative planning and upon the fact that operations aim soon to reach four kilometres — and beyond — into the depths of the earth, deeper by far than man has dug.

While Western Deep Levels is acknowledged as the deepest mine in the world and technologically is regarded as the most advanced mining enterprise of today, this is not to imply that it is the largest — or richest — gold-mining operation in the world, or even in South Africa. For example, as the Guinness Book of Records tells, Driefontein Consolidated, adjacent to Western Deep Levels' area of operations, is probably now the world's richest gold-mining venture, though is not the largest in size. That record belongs to Anglo's Vaal Reefs mine.

But in assessing size and wealth, what criteria does one employ? Perhaps the size of its work-force; perhaps the basic richness of the ore; perhaps its relative profitability. What criteria, indeed? Many come to mind.

To such questions, the South African Year Book of 1986 gives authoritative answers. "Judged on milling rate," it states, "Vaal Reefs is the largest gold mine, having milled 9,5-million tons of ore in 1984 and recovered 81,9 tons of gold." But, the Year Book goes on, "Driefontein Consolidated, formed through the merging of East and West Driefontein mines, is first on a total production basis, having thus far recovered 2 102 tons of gold." These are claimed as world records. In recent years, however, some of these figures would seem to have been surpassed. Freegold, an amalgamation of Anglo mines in the northern Orange Free State, now produces more than a hundred tons of gold a year.

High in the big league of the world's gold producers is Muruntau, situated in a remote corner of the Soviet Union. Like Western Deeps a relatively young mine, it is estimated to produce at least 80 tons of gold a year. Actual figures are kept secret, however. The mine itself is a vast open-cast pit that is both broad and deep, though not yet reaching as far down as Kimberley's famous "Big Hole" where men once delved for diamonds in South Africa.

In the region where Western Deeps is situated, the plains of the Western Transvaal are broken by a series of low parallel ridges running

Gateway: Number 1 Shaft complex.

roughly from east to west. One, the Westwitsrand, bisects Western Deeps' property. The northern sector is shared by Number 2 and Number 3 Shafts, separated by some two and a half kilometres. But for almost a quarter of a century the shallow valley over the hill remained untouched, its rural look contrasting with the active industrial bustle to its north, where, through the years, mine dumps had developed to heights almost matching those of the surrounding ridges.

The ultimate exploitation of the ground to the south had always been inevitable. Initial plans had envisaged more than two shafts for Western Deeps, but development waited for the time when financial and other practical considerations were ready to meet the enormous costs of such an extension. As the years passed, people wondered more and more when a new shaft would take its place in the empty valley. Test drilling had indicated promising prospects there, though the slant of the gold-bearing reefs, dipping fairly steeply towards the south, meant that man would have to delve more deeply there to reach them.

In February 1980 the Company announced its intention of immediately sinking an additional shaft system to exploit the southern portion of the lease area and, next to the shaft, also to construct a plant capable of handling 160 000 tons of ore a month. In his report for 1979 the Chairman, Gerald Langton, described it as "an exciting new development which profoundly affects the future of our mine." He went on to tell shareholders at the annual general meeting that the proposal included the mining of the Ventersdorp Contact Reef in a large area south of the lease boundary where the mineral rights are held by Western Ultra Deep Levels Limited. He did not foresee any problems, however, in acquiring the necessary rights there and in concluding an agreement satisfactory to both parties, though "the final decision to proceed with the project has to await agreement between the two companies regarding the exploitation of this area."

This was the genesis of the Number 1 Shaft.

Initial expenditure of R9,8-million was approved for its development. Without delay work began on the new project which came into production in 1986, ahead of schedule. It was initially expected to be finally completed by 1992 at a cost of R715-million, by which time, it was estimated, 56-million tons of hoisted ore would have yielded some 460 tons of gold.

The recommendation that this "additional shaft system be sunk in the southern portion of Western Deep Levels lease area" had come from Anglo American's Consulting Engineer, at that time Fred Bayley, who had begun his mining career at Western Deeps in 1962 and, in

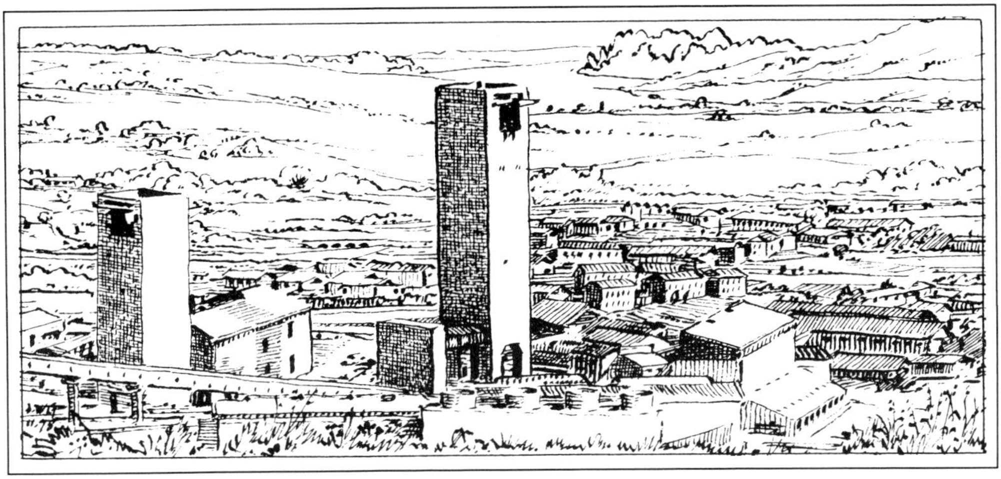

Number 1 Shaft complex, Western Deep Levels mine. Now known as the "South Mine".

spells, had served several years there. As Consulting Engineer he was closely associated with the planning of the new shaft and in such circumstances it was not mere coincidence that shortly afterwards he was appointed Regional General Manager for Western Deeps and the adjacent Elandsrand mine, a post he still holds.

He has thus supervised the development of Western Deeps' Number 1 Shaft "from the drawing board to production".

A twin shaft system was envisaged for this project, and that with the further sinking of a sub-vertical shaft system, access would be gained to a large tonnage of Ventersdorp Contact Reef situated south of the Western Deeps' existing lease area when extended by the incorporation of further ground from Western Ultra Deeps. This development was expected to be completed by 1992, "at which stage," Bayley predicted, "a depth below surface of 3 880 metres will have been reached . . ."

The magnitude of present capital costs, barren of returns over lengthy development times, makes it clear that gone now are the days of the old-fashioned "gold rush" such as California saw with its legendary "forty-niners" almost a century and a half before, and, too, as Australia witnessed later. Nor could the early miners in South Africa have survived the change from alluvial pick and pan workings to modern underground mining.

Gold mining today is an enterprise well beyond the capabilities and financial means of a single individual. True, there are still a few scattered mines privately operated, but those are minuscule both in numbers and in size of operations. Not only does gold mining now require immense investments of finance and sophisticated expertise as well as costly equipment, but it also entails long patient waiting before development begins yielding returns on capital invested.

Sir Lionel Phillips was one of many who, in the early days of the gold industry in South Africa, stressed these facts. Even by the 1890s he appreciated that changing circumstances had brought to an end the picturesque, romantic era of the lonely nomad who exploited alluvial and shallow mineral deposits wherever he stumbled upon them. Earlier this century Sir Lionel pointed out that "the cost of extraction escalated" as the depth of mining increased and that with such increases of depth — inevitable as the surface deposits became worked out — "large investments were required several years before production and, therefore, before earnings and profits could begin."

Modern experience certainly underscores Sir Lionel's perception. Early Western Deeps shareholders will probably not disagree!

Number 1 Gold Plant: Process Flow Sheet.

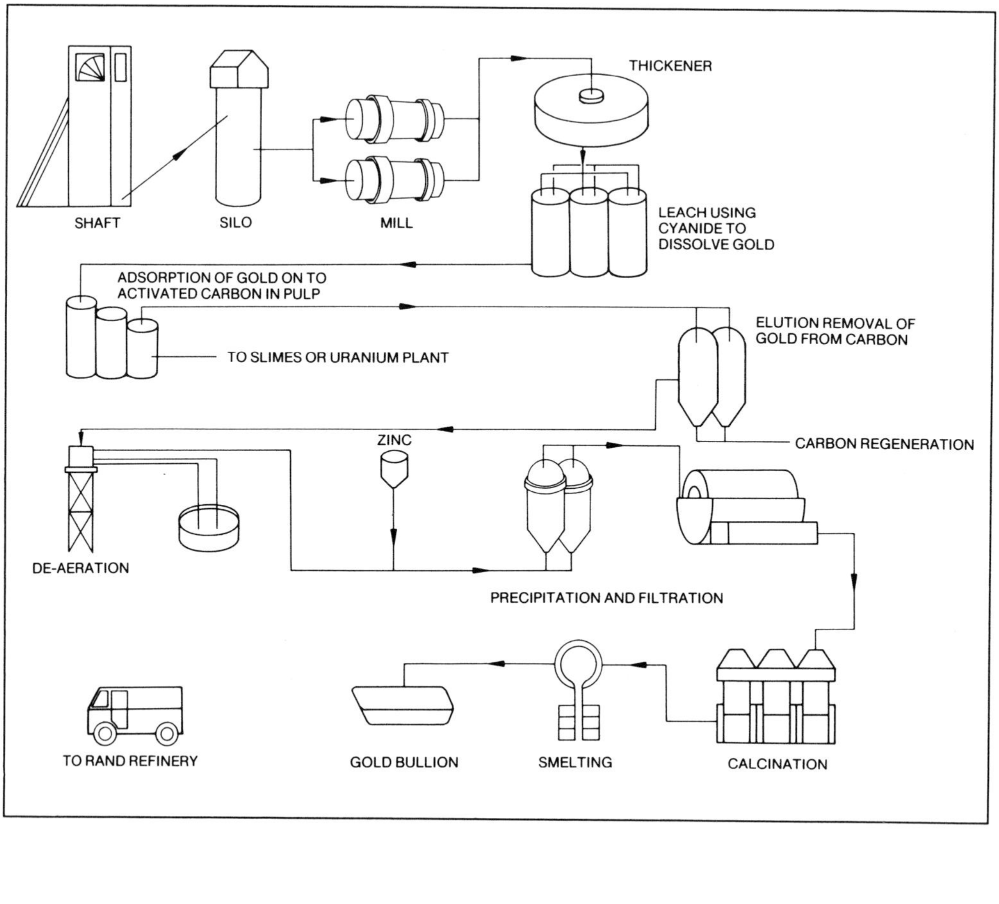

For the most part, the development of the Number 1 Shaft at Western Deeps adhered to plans and schedules. But progress was not without problems and setbacks. In 1982, for example, severe flooding delayed shaft-sinking for eight months. But once production had been reached, all efforts were made to offset the delays and to compensate for the initially financially barren years of development. Soon mining there was yielding about 100 000 tons of rock each month. Of this some 60 000 tons went to the crushers to be processed for gold, and the years of intensive preparatory work began to yield rewarding results. This was an end to the beginning, as it were. Next, fresh plans and schedules were drawn up, envisaging that the earlier estimates could be dramatically increased in the following few years. By 1992 some 120 000 tons of auriferous rock from the shaft will be milled each month. The gold plant had been designed on a modular basis to permit whatever expansion might become necessary.

It was perhaps a coincidence — a happy one, nevertheless — that the date of the ceremonial opening of the Number 1 Shaft at Western Deep Levels in 1986 should coincide with the centenary of the discovery of gold on the Witwatersrand.

By then, Number 1 Shaft had been in production for some months. Its main shaft already reached down beyond 2 500 hard-won metres on the planned journey into the depths of the rock-bound earth. Western Deep Levels itself had been inaugurated twenty-four years earlier and the establishment of this new shaft thus merely marked an extension of the mine's operations.

At Western Deeps the formal inauguration of the Number 1 Shaft was marked by the pouring of the first gold mined and milled from its own operations.

It was a great occasion.

The honour of pouring the first bars of gold was given to Mr Gavin Relly, Chairman of the Anglo American Corporation. It was an appropriate choice not merely because Mr Relly was the head of the parent organisation, but also as an acknowledgement of the considerable assistance rendered to Western Deeps in many ways by Anglo over several years, not least in respect of the development of this latest — and greatest — project. Furthermore, many will recall, Mr Relly himself had been an Alternate Director in the early stages of the Company's growth.

A distinguished group gathered at Number 1 Shaft's smelthouse as the first gold from this section of the mine was poured. It was a proud moment especially for Alan Ramsay and Dick Solms who, as

Project Mine Manager and Metallurgical Manager respectively, had supervised the later stages of the shaft's development prior to production.

Many there, of course, had seen gold poured before. But the romance never palls. There is always a magic — and emotion — in watching the crucible yield its fiery, molten stream to the waiting moulds. The splendour of the yellow metal excites the imagination and stirs the mind to wonder at the toil as well as the adventure involved in its extraction, grain by tiny grain, from the depths of the unforgiving rock.

A standard modern gold bar weighs from 28 to 30 kilograms. That can be reckoned to have taken the crushing of between 4 000 and 5 000 tons of mined rock to produce.

In his address, Mr Relly said that, however important the opening of a new shaft at a gold mine might be seen to be, this opening had far greater significance than just the inauguration of another section of a mine.

"From the beginnings of its history," he went on, "Western Deep Levels has been at the forefront of technical endeavour. The advances achieved here have truly leaped beyond the bounds and frontiers of conventional technology. Many notable firsts have been recorded during the development of this shaft and I can think of very few projects where the interaction between numerous different teams has been so essential to the success of a project as it has been here."

Pointing out that the shaft had come into production in a little more than six years from the time that site clearance had begun, instead of the ten years allowed for in planning, he emphasised how much of that valuable time-saving was due to cohesive and outstanding team work, to close understanding and commitment by all who were involved. This had been seen at all levels. While the engineers, for example, were discussing some of the substantial challenges facing them, workers were getting together and debating practical ways of improving hostel design. This had resulted in better accommodation and housing layout.

Mr Relly praised the extent and degree of management-worker collaboration in the project and stressed that the co-operation at Western Deeps had been a splendid example to everyone.

While he envisaged that mining at this shaft would take place initially at depths of between 2 000 and 3 000 metres, the Number 1 Shaft was unique in that, when it became fully operational, mining would take place at depths approaching four kilometres. "This is almost a journey to the centre of the Earth," he said, "the implications of which would have been the envy of Jules Verne."

Since Western Deep Levels would be the first mine in the world to reach to such depths, inevitably along the way it would have to pioneer new techniques. So far, means had always been developed to meet fresh problems as they had arisen, and success in solving them had bred confidence in the technical skills of the workers to cope with whatever further problems and difficulties might arise.

Clearly the two major obstacles to be overcome in the drive down to working levels at 4 000 metres would be those of intensifying rock pressures — already a headache — and of high temperatures which grow more and more acute the deeper man probes.

Mr Relly also touched upon the considerable changes to be expected in South Africa's developing society in the years ahead, and suggested that the success of the co-operation between management and workers at Western Deeps gave hope for the future of the country. The prospect of change was not something to be feared, he said. Pointing out that the cost of this project was enormous, he said that the capital outlay showed confidence not only in the continuing international demand for gold, but also, "as it is expected that Number 1 Shaft will be mining well into the year 2030, it also reflects confidence in the future of the country."

That optimism was later reinforced by Mr Harry Oppenheimer addressing the guests at the luncheon celebrating the occasion.

"When one sees what is happening here, technically, and what is happening here socially, it gives one — or gives me anyhow and I hope it gives you — great hope for South Africa, in spite of all the troubles which are going on," Mr Oppenheimer said.

The luncheon was held in a large marquee on a grassy terrace on the Westwitsrand, a stone's throw away from the main offices administering the South region of the sprawling Western Deep Levels mine. To the east of Western Deeps' property, the Westwitsrand and the Gatsrand come together like the segments of a chicken's wishbone. The broad valley in between is shared with two other well-known, older and successful deep-level mines, Blyvooruitzicht and Driefontein Consolidated, both contiguous with WDL on its northern boundaries. Situated as they are where the southward dipping auriferous reefs lie nearer the surface, these two mines do not have to delve as deeply for their gold as Western Deeps has to do.

From the surface of a mine property, it is not easy to envisage the boundaries of the underground working area, to know where mining limits begin and end. Natural features on the ground, such as hills and roads, for example, which would normally provide appropriate boundaries to any property, do not necessarily coincide — on a mine — with the declared limits of the working area beneath.

Certainly this is so at Western Deep Levels.

Here, to the south and to the east, roadways would seem to provide natural boundaries. But they are misleading. Underground, the mine stretches well to the south of the busy highway that links Johannesburg to the east with Potchefstroom, 60 kilometres to the west. Similarly, on the eastern side, the road from Carletonville — which winds gently over the grassy ridge before crossing the highway and running on to Fochville — is not the boundary that it appears to be. Underground operations extend unseen some distance up the valley to the east.

From the ridge where the luncheon marquee stood on that day, the view to the south is an impressive one, beautiful in a typically South African highveld way. Beyond the koppies across the valley, one looks out upon a broad plain that stretches on and on to low hills far away on the distant horizon. On that morning the hills were pale mauve in the sunny, dusty and smoky haze of a Transvaal mid-winter's day. The veld was made a chequer-board of black and gold by the copper-colours of the tawny dry grass interspersed with large irregular patches recently burned by farmers fearful of uncontrolled fires in the tinder-dry grazing paddocks thirsty for the early spring rains.

Typical highveld agricultural countryside — "earth from which bread comes": as once also was the Western Deeps property, with mielie fields and grazing lands: wide expanses for crops, plentiful grazing for cattle and sheep. At the luncheon, guests with sharp eyes might have seen the outlines of drilling derricks — dwarfed by the sprawl of the landscape — patiently probing for the new gold mines of the years ahead. The bunting above the marquee fluttered in the breeze, while proudly over the office of the South region the South African flag flew beside the Western Deep Levels' standard — on a royal blue ground a golden falcon: the ancient symbol of gold mining.

With luncheon drawing to a close, Humphrey Hobday, recently entrusted with the responsibilities of managing this prestigious Number 1 Shaft, called upon Peter Gush, Chairman of Western Deeps, to introduce Mr Harry Oppenheimer, the distinguished former Chairman of WDL and also of both the Anglo American Corporation and De Beers Consolidated Diamonds.

For Mr Oppenheimer the occasion must have revived many memories.

It was appropriate that Peter Gush should invite Mr Oppenheimer to address the gathering. A Rhodes Scholar and Oxford rugby Blue, Gush was not only Chairman of the Western Deep Levels Mining Company at that time, but also Chairman of Anglo's Gold and Uranium Division. Furthermore, in that centenary year of 1986 he was President of the South African Chamber of Mines. This led Mr Oppenheimer to say that "he is not only the head of the Anglo American gold mining division, but at the head of the mining industry of South Africa."

"There is a very long history which led up to the ceremony we are taking part in today," Mr Oppenheimer told guests, and went on to mention some of the significant developments between the beginnings of the mine and the opening of this new shaft.

Touching deftly upon salient points in the conception and the growth of the Western Deep Levels Gold Mining Company, Mr Oppenheimer paid tribute to the Gold Fields of South Africa, whose interest in the western area — 50 kilometres from the historic first gold strike at Langlaagte — had led to its exploration and development. It is a fascinating story. First to discover the "vanishing reef", Gold Fields offered other South African major mining houses participation in the exploitation of the reef. It was an offer which Anglo American accepted and which led in due course to Anglo's developing its own operations in the region. It was an attractive gamble, "but," Mr Oppenheimer confessed, "we didn't realise how or what this was going to lead to."

Up till then, he went on, mining at the great depths demanded by the reef in the area had been considered to be beyond man's ability. Engineers had felt, however, that there were technical developments in the offing which would make such deep mining possible — and profitable — provided the ore grades were reasonably high. Thus prompted, the Corporation had begun a closer examination of the West Wits area. It was Dr Austen Bancroft, "a great geologist in our Group", who, encouraged by Sir Ernest Oppenheimer, urged further developments and "picked on this area as the one where there was the best chance of developing mining at depths which were unheard of in those days."

The degree of this uncertainty in the future of such mining was underscored by the fact that when Anglo American offered a 10 per cent share in the venture to Newmont, the great American mining company, the offer was turned down. It was considered too risky.

Faith and hope were the spurs to an enterprise which called for courage. It was Sir Ernest's faith in the judgement of his advisers, however, which eventually persuaded him to go ahead with the project, though in the face of no little scepticism.

With nostalgia and, understandably, some personal emotion too, Mr Oppenheimer recalled the ceremonial beginnings of the mine. "It always pleases me to think back," he reflected, "to 1957 — it's a very long time ago — when my father, who was an old man, came and performed the ceremonial setting going of the pre-grouting drills in July 1957 — only a few months before he died."

The boldness of the project inevitably brought with it worries and problems. They served to sharpen the edge of challenge. Looking back over the past thirty years, Mr Oppenheimer recalled the enormous pleasure of now seeing the development of this exciting and successful venture.

He outlined the social work instituted — and continuing — at the mine and mentioned the launching of the Wedela village. This he hoped would soon turn from a village into a township of considerable size. He said how deeply he felt that the combination of technical and social achievement gave great hope for the future of South Africa, in spite of the manifold problems of the present.

"It shows a different approach to African problems from the approach which has been generally adopted, often very disastrously, in the African continent," he said. People had thought of progress and change purely in the political field, leaving questions of economic and social development to be settled later. He stressed that the various aspects of change ought to be tackled simultaneously, or at least that social and economic problems be tackled before drastic changes in the distribution of political power. What was being done at Western Deeps was, he felt, the "sort of thing which will make the changes in political power — which undoubtedly have to come — more successful, more peaceful, more reasonable, more desirable in every way than they've been elsewhere in Africa."

The word "Wedela" is derived from a combination of the names of two of Anglo's gold-mines, Western Deep Levels and Elandsrand, situated alongside one another. The township is a model of its kind, situated in attractive rolling countryside between the two mines. It caters for married black staff at both.

The housing of black mining staff had long been a matter close to Harry Oppenheimer's heart, and the creation of Wedela, which he had formally opened at the end of 1978, must have given him immense personal satisfaction. It was the realisation of a plea he had eloquently made in the House of Assembly more than twenty-five years earlier, calling for adequate housing for married black staff on the mines, a proposal which was opposed by the government.

Hostel at Number 1 Shaft: the Administration building in the foreground and the Information Centre behind.

Now the wheel had turned full circle: here, more than a quarter of a century later, Wedela was the visible manifestation of Mr Oppenheimer's thinking.

The media, especially South Africa's mining press, gave considerable publicity to the inauguration of this new shaft, describing, in one instance, Western Deeps as a "miracle mine", while another report called it a "mega-mine".

Guests at the opening ceremony were handed two pamphlets. One, prepared by Dick Solms and Alan Ramsay, joint Project Managers at the Number 1 Shaft, provided interesting technical information — much of it in the form of statistics and figures.

Inter alia, the technical brochure described the configuration of the "mega shaft" and its "unique features"; combining a production shaft and a separate service shaft, and went on to give details about the Blair multi-rope winders capable of hoisting 400 000 tons of rock per month from a depth of 2 300 metres. The depth of mining operations made sub-vertical shafts necessary below 83 Level. Also installed were winders "capable of hoisting two 150-men cages per trip."

Diesel-fuelled load-haul-dumpers, tested some years earlier at Number 2 Shaft, are used for all stope production at Number 1 Shaft, the first South African gold-mine solely to use this trackless transport equipment.

Shaft sinking operations had seen a number of engineering innovations: the largest kibble ropes in the world, for example, and the use of the world's heaviest and longest winding rope, almost 15 kilometres in length! In due course, the kibbles (buckets used for hoisting rock) were replaced by the more usual skips.

The main shaft at the new complex is "a circular monolithic concrete-lined" construction with a diameter of 9,8 metres and reaching to a depth of 2 374 metres. The headgear above it rears to a height of 83 metres.

To reduce the high temperatures at the working face, vast quantities of chilled water and ice slurry are pumped underground, necessitating huge refrigeration plants on surface as well as massive pumping equipment underground.

The other brochure was the reprint of an article by Hugh Elliot, Chief Geologist at Western Deeps, which had appeared in *Optima*, a quarterly magazine issued jointly by Anglo and De Beers for their shareholders. The editor had given the article the title "Down towards the Styx".

In different ways, each souvenir booklet depicted on its cover the Western Deep Levels' crest — a stylised version of the falcon, similar to the so-called soap-stone bird of the Great Zimbabwe and now incorporated in the national crest of Zimbabwe. Some have criticised the possible plagiarism, but in fact Western Deep Levels — with numbers of miners who had also worked in what once was Rhodesia — had adopted that symbol on its crest before the falcon became Zimbabwe's national emblem. There it was inspired by carvings at the historic ruins of what was believed once to have been a great city and a rich trading centre, known for its gold exports. For Western Deeps the inspiration came from the falcon's historic link with gold.

Many ancient writings tell of the connection between the falcon, on the one hand, and gold and mining, on the other. The bird has been depicted in rock engravings near gold-mines in ancient Egypt and the symbol has been traced also to West Africa. In the Old Testament too the falcon is linked with mining and gems.

John Shilling, the first Managing Director of Western Deeps and, later, its Chairman, did much research into the mythology of the falcon's sacred role in gold mining. This led Douglas Hoffe, when he was Managing Director (1963 - 1964), to embody the bird in a crest for

Western Deeps. The significance of the falcon, he feels, "... captures the mystery and magnificence of gold, a symbol of wealth from the dawn of time."

Initially eight shafts had been envisaged for Western Deep Levels. This Number 1 Shaft was the third to be brought into production. By its size it is a mine in itself. Costs have been enormous and it now seems unnecessary that those original plans will be implemented. Improved technology may make it possible to exploit the lease area through existing shafts. There may perhaps be one more new shaft — at most, possibly two — developed in the years ahead, but more than that are certainly unlikely. It is currently planned that all ore mined in future will be hoisted through the existing three shafts, using an expanding network of horizontal tunnels on the auriferous reefs underground.

But at present such developments remain conjectural, lying beyond the horizons of time, and the implementation of plans, however carefully and thoroughly prepared, is always subject — not least — to the whims of nature, as these pages tell.

Western Deeps' Number 1 Shaft, itself heir to a significant number of innovative developments, is set to expand the existing treasury of mining technology and knowledge. With its feet in the past, as it were, its eyes are fixed on the future, ready to make further history in a region itself historically significant.

CHAPTER TWO

Where did the Gold come from?

The centre of modern Johannesburg is much like the downtown business area of almost any other First World city anywhere. The same styles of high-rise office blocks reach towards the sky; the shop windows display the fashions and goods that could come, season for season, from other sophisticated cities on other continents. The broad streets, noisy with traffic, are concrete canyons which funnel the wind and hold back the sunshine from the busy pavements.

From the higher ground to the north of central Johannesburg one catches glimpses, between the buildings, of the old mine dumps which characterise the western side of the city. Pale ochre in colour, the sides of these man-made hills are darkened by a thin covering of straggling grass struggling to survive in the chemical-tainted earth, pulverised rock which once came from deep underground when, not very long ago, gold was mined there.

A little further westward and to the north, where the city's neat commercial centre yields increasingly to industry, one sees here and there the abandoned headgears, now derelict and rusting, of gold-mines which made Johannesburg rich and famous and great: leviathans now out of time and out of place: memorials to a past generation of pioneers.

Today they act as signposts to the rich contemporary gold-fields of the western Transvaal. Along the highways which run towards the setting sun are some of South Africa's best-known and richest gold-mines.

Here they straddle what geologists have named the Witwatersrand Basin which lies south of modern Johannesburg, stretching from slightly south-east of the city far to the south-west and south. This,

Main entrance to Western Deep Levels mine property. Number 3 Shaft in background.

millions of years ago, was a mighty cratonic depression which, with the changing climate, filled with water to become a large inland sea. Now, beneath the thick lava flows and sediments, it contains the largest deposits of gold known to man. Underlying a large part of the Orange Free State and a small part of the Transvaal beneath the highveld, the Witwatersrand Basin has been described as "one of the wonders of the world of minerals".

The fabulous richness of its contents at once stamps the Basin as unique. Despite all that has already been mined, geologists are confident that its reefs, deep lying, will yet yield vast quantities of gold. Nature guards the prize. Its richness challenges man.

Science looks to the future with confident assurance. But, ironically perhaps, the past seems less clear. What is less certainly known is how the gold came to be there. Theories are plentiful both about the source and the geological timing, but none is certain and none without dispute. In a notable understatement the Great Soviet Encyclopedia says that "The problem of the genesis of the Witwatersrand is still debated." An Anglo American geologist has said that theories "change between Thursday and Sunday!"

Since, as we now know, much of the gold lies at great depths — in this particular area in undulating strata between 800 and 6 000 metres underground — the Basin must have been extremely deep when some geological phenomenon laid down the first deposits. Now filled with the sediment of an infinite age, the Basin is reckoned to be 12 000 metres deep, a vast chasm created in that Pre-Cambrian era between

2 300- and 2 800-million years ago and filled with the detritus of the ages which followed.

And since the gold (and other minerals) in this rich area to the west of Johannesburg is not concentrated in a single stratum, but is spread over a varying number of levels, it must be assumed that there was more than merely one era during which gold was brought from other sources and deposited — like a multiple sandwich — between layers of other rocks and detrital material, as diagrams in this book illustrate.

It is generally accepted that our world began some 4 550-million years ago. What gave rise to the formation of the earth and of the moon and the planets which form part of the galaxy is a complex study, and one which, however fascinating, really has no place here. But it is

Outline map of South Africa, showing position of the "Golden Arc" — the rich gold-bearing reef of the Transvaal and the Orange Free State.

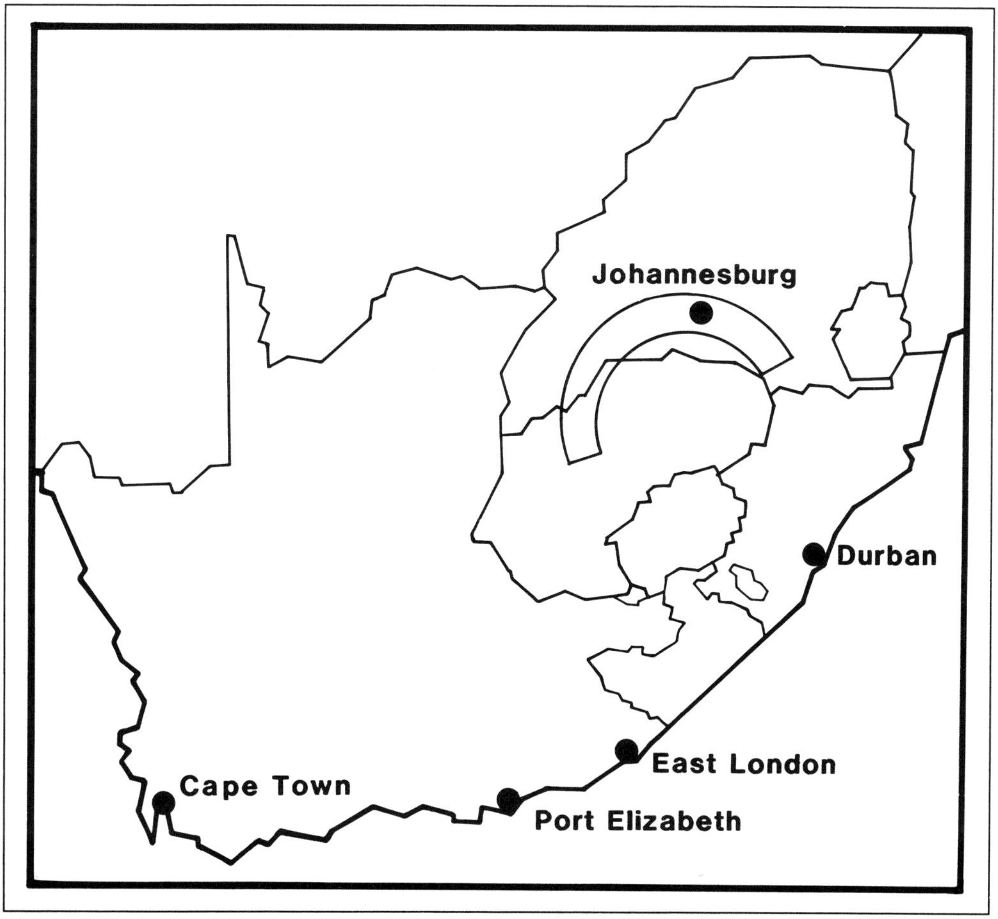

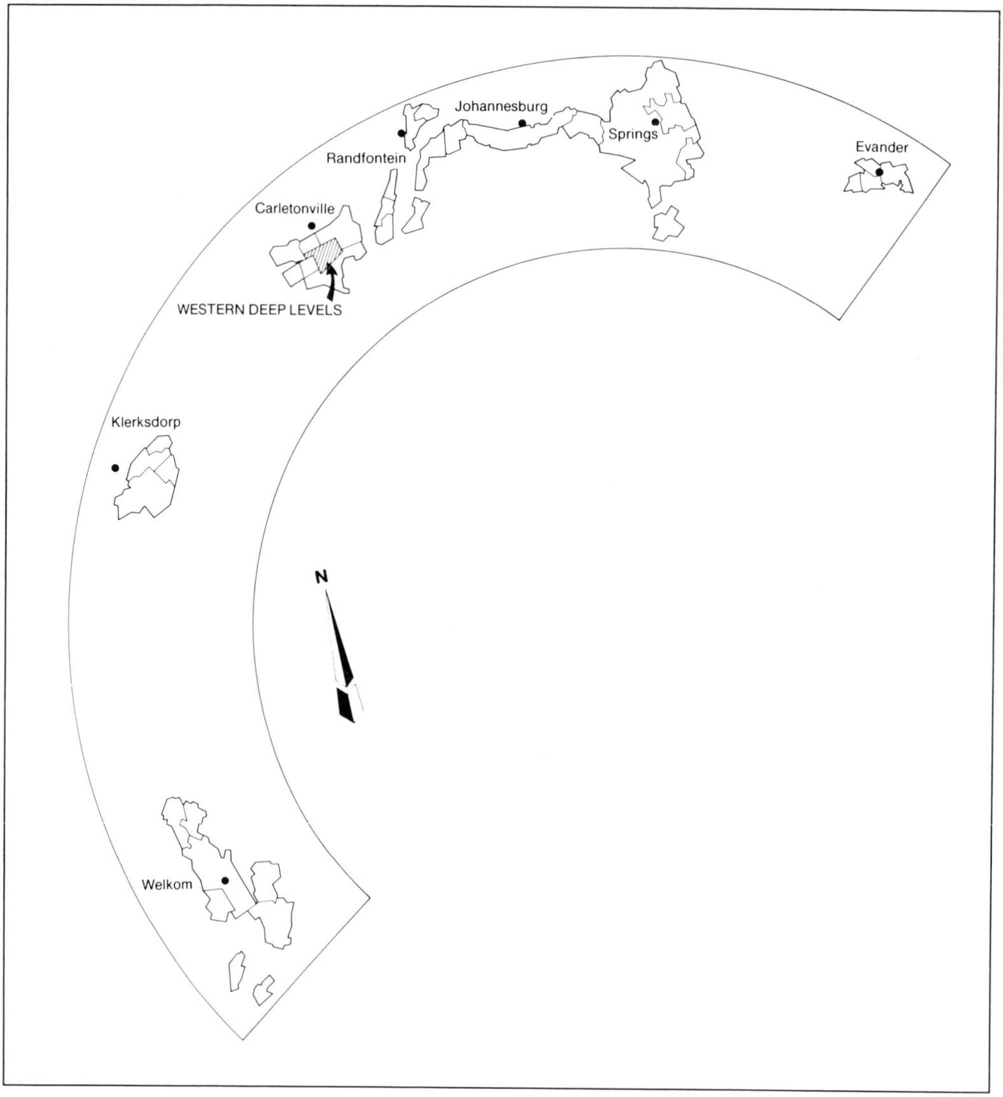

Enlargement of "Golden Arc", showing positions of major mining areas.

interesting to note that the first rocks in the earth's crust are thought to be not older than 3 980-million years, so that we are left to wonder what the earth's crust consisted of in those initial 600-million years. Its surface was largely covered by lava, spewed from a profusion of fiery volcanoes scattered across the bleak, rugged landscape.

At that time, scientists believe, the molten mass of the earth was cooling, from the surface inwards. As its crust began to solidify, its skin wrinkled under the influence of the heat within and below it and these massive wrinkles became the world's first mountains and valleys.

Number 1 Shaft, looking south from the Westwitsrand.

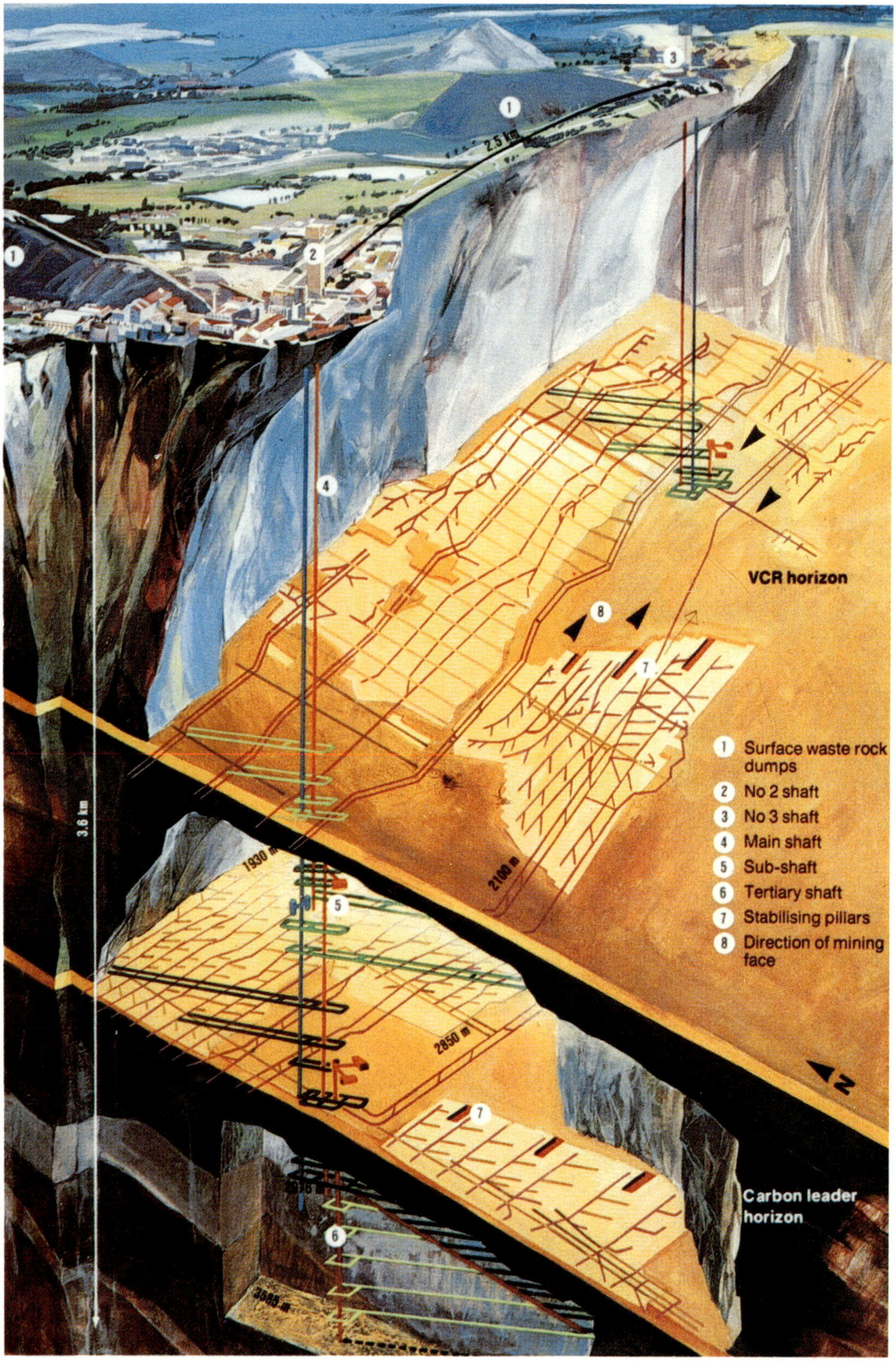

Diagram: Cross-section of Numbers 2 and 3 Shafts at Western Deep Levels, showing positions of Ventersdorp Contact Reef and Carbon, and illustrating types of shafts in use.

Number 3 Shaft with Number 2 Shaft in the background.

Miners working underground.

Pump station on Level 121 of Number 2 Shaft.

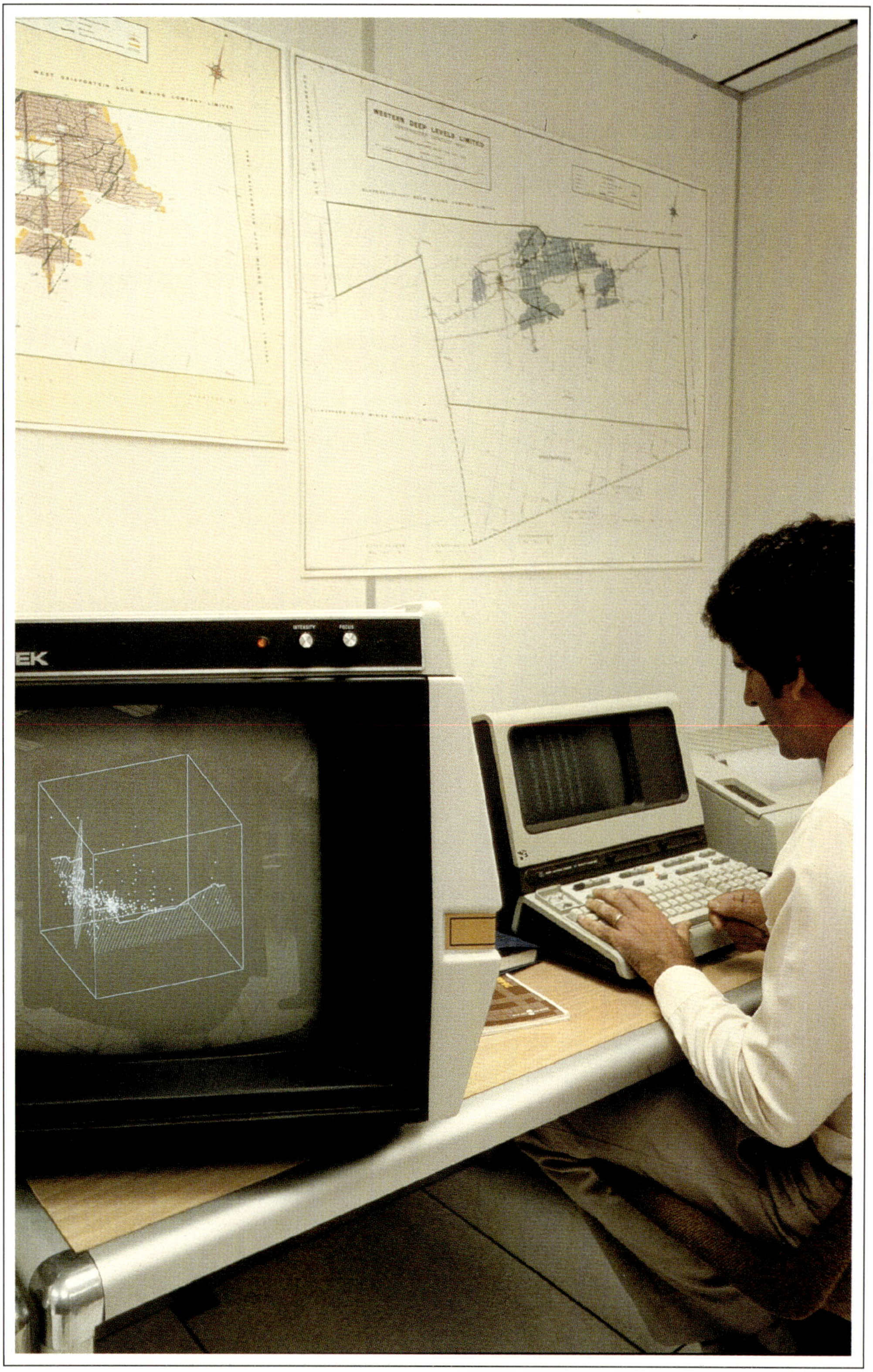

Seismic Laboratory, Western Deep Levels.

Refrigeration Plant: Number 3 Shaft.

Western Deep Levels athletes training at the mine's stadium.

Geologists earlier believed that not all of those first — "greenstone" — mountains have been eroded. Some still exist and are to be found scattered in some parts of the world, including South Africa at Barberton. They have been used by modern scientists as bases against which the ages of other geological formations may be measured and assessed.

More recent evidence tends to indicate, however, that what were thought to be primary mountains of rock are, in fact, created from sediments eroded through millions of years from the first mountains.

The oldest of the world's metasediments, reckoned to be about 3 750-million years old, are situated in Greenland. The greenstone rocks at Barberton are younger, being (it is estimated) only about 3 400-million years old.

But those in southern Africa are of particular scientific significance. They represent, it has been said, "the earliest clearly decipherable geological events on the Earth's surface" and, as such, provide valuable clues to the early evolution of our biosphere. Extensive use has been made of these geological phenomena in the study of developments of the early atmosphere as well as the structure of the earth.

Apart from these greenstone belts in South Africa, mainly in the Transvaal, there are other deposits which host gold.

Early theories about the origins of the gold of the Witwatersrand and adjacent areas of the southern Transvaal postulated that, in the course of hundreds of millions of years, the metal had been carried from the greenstone-hosted lodes in those early-born mountains of the Barberton complex.

Again, however, more recent studies contradict that and indicate that the gold almost certainly did not come from so far away. Indeed, contemporary geologists consider that the metal in such quantities could not have been carried by natural forces over distances of several hundred kilometres. As a South African geologist has recently firmly stated, "There is no convincing evidence that large amounts of gold and uranium were transported in solution over distances exceeding a few centimetres."

It is not mere human curiosity which prompts continuing research into this fascinating question about the source of the Witwatersrand gold. The need for research is a very real and practical one, for such geological information is essential to the future of the mining industry.

Research thus continues, accelerated by the growing economic importance of South Africa's gold mining industry and by the need for fresh resources to satisfy the world's insatiable demands.

Refinements in the development of scientific equipment and in human knowledge have thrown fresh light on the source of the Witwatersrand's gold. Theories about origins in the greenstone lodes of the northern Transvaal are now being discounted by stronger evidence of its origins elsewhere.

For one thing, no greenstone lodes in southern Africa — or, indeed, anywhere else in the world — contain as much gold as the ores of the Witwatersrand Basin. The Witwatersrand is about thirty times richer than the greenstone ores of the north-eastern Transvaal, emphatically suggesting different origins for each. In short, those greenstone lodes around Barberton were clearly never rich enough to provide the prodigious amounts of gold now being mined on the Witwatersrand. Compared too with the richest and largest greenstone gold-producing area in the world, the Abitibi Belt in Canada, the gold content of the Witwatersrand is infinitely larger. This leads to the belief that the source of the Witwatersrand gold was wholly different and did not lie in those formations, even though there are certain pertinent similarities in the physical characteristics of the greenstone-gold and that of the Witwatersrand.

Studies continue: additional information fuels the debate. In essence, much of the scientific discussion is concerned with the ways in which the gold became deposited in the Basin, rather than where the mineral came from. Today's economics, however, demand that evidence provided by known deposits be utilised to track down the original source of the gold — and, equally, to see whether the source has perhaps fed other deposits which might be exploited.

There is growing evidence to indicate that the Witwatersrand obtained its gold at about the same time — in geological terms — as grains of the precious metal were deposited and absorbed by the ancient greenstone rocks scattered about the earth's crust, between 2 600- and 2 900-million years ago. Why the gold, then brought from the earth's fiery core in molten form or in solution, was concentrated in limited areas and not distributed more widely and evenly about the earth presents scientists with yet another complex riddle.

Hugh Elliot, Anglo American's Chief Geologist for the West Region, talks about these momentous geological events as if they had happened last week, and deals with a million years as if such time spells were read from a wrist-watch.

He points to rock samples scattered on shelves and tables in his office at Western Deeps and picks up drill core samples of strata laid down in the Basin hundreds-of-millions of years ago.

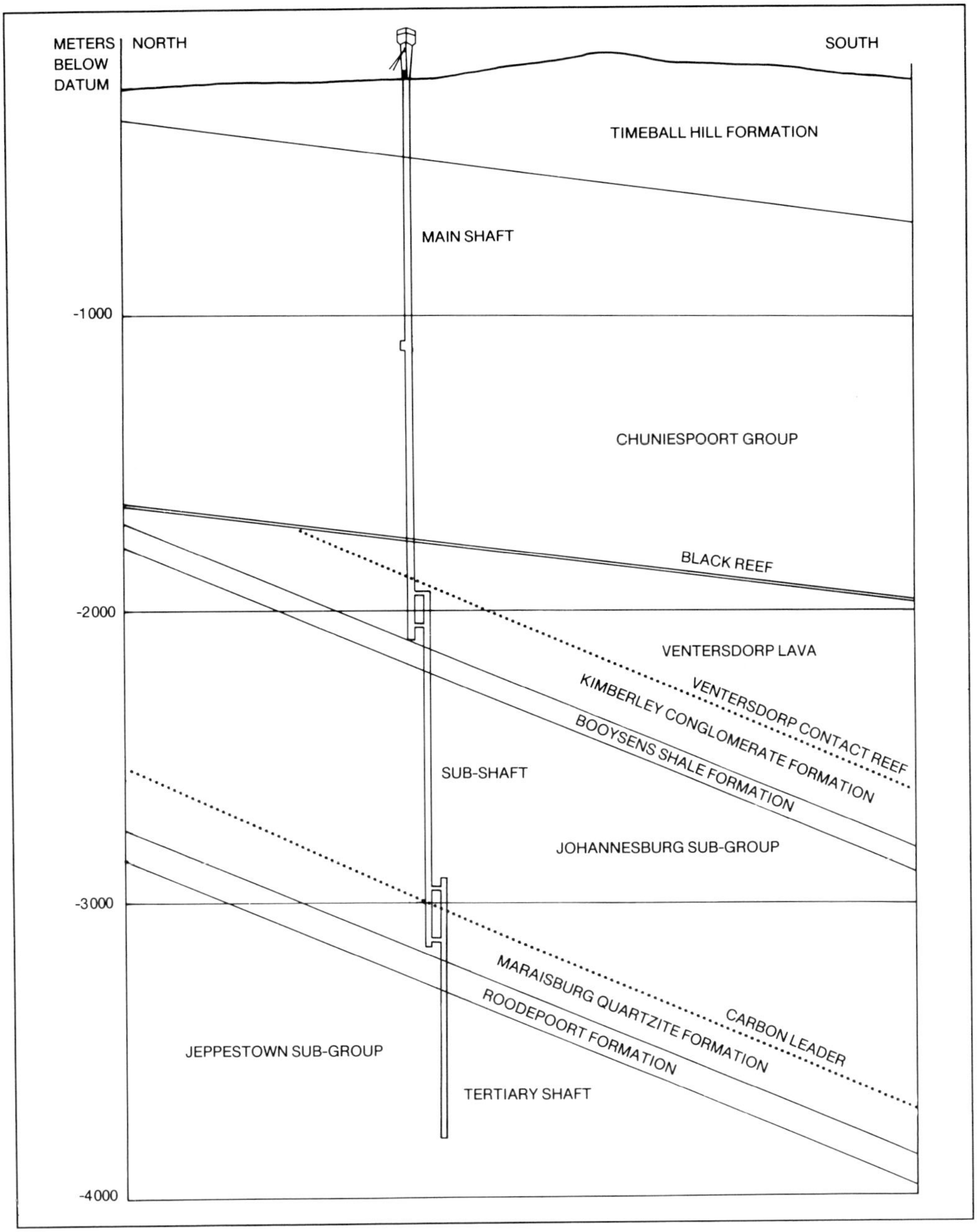

Geological structure in the vicinity of Number 2 Shaft.

One theory maintains that the gold was deposited in the Witwatersrand Basin about 2 750-million years ago by streams and rivers carrying eroded debris containing gold which was gradually deposited in placer formations on the bed of this deep inland sea.

Huge domes of granite rimmed the vast lake, mountains embedded in the earth's crust, as indeed they still are today: massive rocky mountains, some visible above the surface. These contained gold which had either been embedded in the molten rock before it had cooled, or had been introduced into the solidifying rock in solution which had then crystallised out in quartz veins. It is now considered more likely that these adjacent domes were the source of the Witwatersrand's gold, rather than the greenstone lodes further to the north.

But while uncertainty prevails about the source of the detritus deposited into the Witwatersrand Basin, geologists are now able to piece together the history of the evolution of the various strata. Not all are aurifeous. The gold in that region is located in only three reefs, of which the Ventersdorp Contact Reef and the Carbon Leader are mined at Western Deeps.

Of the two, the Carbon Leader is the more important, here found between 2 340 and 4 500 metres below the surface. The gold is contained mainly between small pebbles established in a quartzite matrix, lying about 30 centimetres thick, with its base unconformable upon the irregularities of the bed on which it rests. Gold exists too in veined bands contained within the conglomerates and quartzites. In these forms the host reefs tend to lie in wide channels that, on the Western Deeps mining area, run north-south and often, to their depths of almost two metres, are filled with the auriferous conglomerates.

The presence of carbon in the body of the reef gives the Carbon Leader its name: sometimes disseminated as tiny grains — flyspecks — within the conglomerate, but more often present as a laterally continuous seam between one and seven millimetres thick, composed of fibres containing organic structures similar to types of algae and modern lichens. Frequently these fibres are coated with gold.

The Carbon Leader lies in a broad table. Its characteristics indicate strongly that it is a fluvial deposit spread in the course of several pluvial periods with possible subsequent modification by marine agencies.

The stratigraphical position of the Ventersdorp Contact Reef indicates that this formation was developed much later than the Carbon Leader in the geological sequence of the earth's formation. It too derives from gold-bearing fluvial deposits that flowed from the north into the basin, which, by then, had filled up considerably and, under the weight of the accumulating detritus, had subsided very slowly, its base developing a southward tilt. The margins on which the fluvial fans were active canted very little. The severe tilting of the reefs — to the extent of 21 to 24 degrees as we see today — was an ongoing process which halted only when the basin had been filled with sediment and, later, covered with lava.

The Ventersdorp Contact Reef derived its name from the fact that it lies at the lower contact of the Ventersdorp Supergroup (of lavas) with the Witwatersrand rocks. The type section of these lavas is near the little town of Ventersdorp, which took its name from the original owner of the farm on which the settlement was founded.

But as rich as the Transvaal is in this precious metal, the Witwatersrand by no means has the monopoly of Africa's gold. Nor, indeed, is gold exclusive to Africa.

Nor by any means is gold only a modern treasure. Its value is not merely a whim of recent centuries. Gold has been sought and valued by man for thousands of years. Almost certainly it was the first metal known to man. Its use has not been limited to any single civilisation, nor has its value been a fashion of any one tribe, race or nationality. Indeed, it is the very universality of its acceptance which has not only maintained its value, but indeed has extended the scope of its use across the world through many millennia into modern times. We are not sure when gold was first used. Archaeologists think it may possibly have been worked even as far back as later Neolithic times, but there is no certainty about that.

What is certain, however, is that at least very soon after Neolithic times alluvial gold was collected from the rivers of Armenia and worked there for several thousand years until the deposits petered out shortly before the time of Christ.

The well-known ancient Greek myth of Jason and the Golden Fleece provides a picturesque clue to one method used by the ancients to collect gold from the rivers and streams, simply by laying the skin of a slaughtered sheep in the bed of an auriferous stream and allowing the waterborne specks of gold to adhere to the natural oils of the fleece. It was then dried in the sunshine, or, as Jason found when stealing Aeetes' fleece, hung from a tree to be dried in the wind before the particles of gold were scraped from their woolly base.

This manner of recovering the metal has not vanished with the passing years. In modern times early prospectors in South Africa, and elsewhere, were known to use similar methods for collecting gold from streams.

Certainly gold was known to man in Egypt more than 6 000 years ago. It was there that some of the most ancient discoveries were made. In two particular areas — Upper Egypt and Nubia — gold was mined extensively from about 4 000 years before Christ. Fresh exploration is now seeking to resuscitate those ancient mines which have been abandoned for about 2 000 years.

The historical gold deposits of Upper Egypt were water-laid, the results of gold scoured from auriferous rock through millions of years by an extensive river system which now no longer exists. Prospectors first took the gold from alluvial deposits by panning in much the same way that prospectors in the nineteenth century won gold from rivers and creeks in the gold rushes in America, Australia, New Zealand and South Africa. In the course of the fourth millennium BC, however, the Egyptians developed underground mining and, as remains still indicate, their engineering skills led to the creation of shafts of remarkable depth and to tunnels of surprising length.

In that ancient world, not only did the techniques of mining itself become remarkably well developed, but the knowledge and methods of refining the gold led also to standards of purity which compare well with those of today. Between 4 000 and 2 000 BC the Egyptians mined prodigious quantities of gold, most of it going to the kings and royal families of those times to be used for the adornment of their temples and tombs. In later years, pillaged and robbed of their wealth, scarcely a single tomb of that era has survived untouched. Today the world mourns not only the disappearance of that gold but, more importantly, also the desecration of the unique art for which it was used.

Egyptian searches for gold not only ranged through the Middle East and into countries along the shores of the Mediterranean and the Red Sea, but also ventured southward to the Indian Ocean and down the eastern shores of Africa as far as the mouth of the Zambezi. This seems to have been the region which they named Punt and which developed into a source of considerable trade in gold, ivory and slaves.

Much of our knowledge of this epoch comes from stone engravings which have survived the ravages both of time and desecration. An interesting hieroglyphic carving in a slab that was once in an Egyptian temple of the fifteenth century BC, shows a galley of Queen Hatshepsut's fleet being loaded by slaves at Punt, watched with curious interest by falcons, traditionally associated with gold and mining.

While considerable debate continues about the location of the wealthy gold centre of Ophir, where King Solomon's mines were located, there is evidence that it probably existed in East Africa and could indeed be the same place as the more ancient land of Punt. That Zimbabwe was once a wealthy source of gold — the refining centre for metal mined in a large surrounding area — there is little doubt. But controversy yet continues as to the age of Zimbabwe's gold industry and whether the place indeed possessed links with the Middle East as long ago as those pre-Christian millennia.

In such circumstances, it is not strange that Africa's riches have captured the attention of writers and explorers through the ages and that the continent's gold has featured so extensively in history as well as in fiction.

Centuries were to elapse before the collapse of overland trading led Europeans to the southern and eastern shores of Africa and the uncharted waters of the Indian Ocean.

With their precious cargoes of spices, silks and brocades and exotic merchandise sought after in Europe, the navigators also brought back accounts of the incredible wealth of Monomotapa, the African potentate, whose kingdom was rich and vast and extended over infinite tracts of the continent, in parts of which gold was mined. Covetous Portuguese eyes looked southward as reports of Africa's wealth reached Europe. That the Kingdom of Monomotapa indeed possessed large quantities of gold seems beyond doubt. Fable was founded in fact. From Sofala in the fifteenth and sixteenth centuries, gold was traded with Europe, while much also went to the established markets of India and the Middle East.

Records of early Portuguese traders describe how mines in East Africa were dug to a depth of "a good stone's throw" and how from the exposed veins the workers took earth that was mixed with gold. Accounts relate how the mixture was then heated in a pot to extract the metal. It was a thriving industry. The natives also collected alluvial gold and selected veined rocks which they crushed to a powder that was then washed to extract the precious residue.

There is some confusion about the actual boundaries of the vast Kingdom of Monomotapa. In the sixteenth century the Portuguese seem to have had a shrewder idea of its dimensions and location than some later cartographers. In their contacts with the interior the Portuguese used Sofala and other neighbouring ports of the Indian Ocean, as Muslim traders had done for centuries. These ports provided the shortest route to the inland capital located, roughly, in the central region of what today is Zimbabwe.

A later map, however, drawn in Paris in 1694 in the decorative style of the time and dedicated to the Duke of Burgundy, shows the Kingdom of Monomotapa stretching from well to the north of modern South Africa to the southern edge of the Karoo, in that area where the great African plateau suddenly ends and the escarpment drops dramatically in a rocky cascade to the coastal plains of the Cape.

With their legacy of contact with the Portuguese, it was not surprising that the early Dutch settlers at the Cape, a century later, were also aware of the great wealth of Monomotapa. It may have been the contemporary thinking on which maps were based that possibly led them to believe the realm of that African king was in fact closer to the Cape than it really was.

A map drawn by the Hollander Jan Huygen van Linschoten similarly shows the kingdom of Monomotapa extending southward from near the Equator to mountains at the southern edge of the Karoo.

In his diaries Jan van Riebeeck at the Cape records his opinion that, were it not for the hostile Kaapmans, who blocked communication between the fort and the rest of Africa, people would "come over land to visit us all the way from Monomotapa ... just as, according to the belief of many, they trade with Angola over land and travel from one to the other. Angola's gold is said to come from Monomotapa ..."

Writing on 10 November 1660, Van Riebeeck tells that "The news having been spread that an expedition would be sent to Monomotapa on the conditions mentioned below, a certain Jan Danckaert of Nijnoven has offered to undertake the leadership of such an expedition of volunteers ..." The volunteers were described as "fine, brave, able-bodied men, who have offered their services ... for this expedition of their own eagerness and inclination .."

In "overcast, misty weather" on 12 November the party "left on their journey to Monomotapa in the name of the Lord", carrying a quaint assortment of provisions and equipment on three pack-oxen. Their expedition came to naught, however.

Van Riebeeck's own searches for silver and gold closer at hand were in vain. In January 1654, less than two years after he had landed, a silver-hued metal was uncovered not far from the Castle, raising hopes of precious metals at the Cape. But, as tests were quick to show, what was thought to be silver unfortunately turned out to be valueless.

It was left to Van Riebeeck's successor to discover valuable deposits of copper in Namaqualand — too remote and isolated then to be exploited, however — and for the peoples of later ages to uncover and develop South Africa's rich mineral treasures.

But if we are to accept the concept of those early French and Dutch cartographers regarding the extent of Monomotapa's great realm — reaching south, as the decorative maps show, to the very foothills of the Cape's Hex River Mountains — then we realise that modern South Africa's gold-mining industry, and Western Deep Levels, must be situated in the very heart of that historical gold-rich kingdom.

CHAPTER THREE

The Advent of Man — and the Coming of Men

The headgears, buildings and paraphernalia of several mines dot the area of the western Transvaal where Western Deep Levels is now making history of its own. All day and throughout the night in the shallow valleys of the region there is the steady bustle of industry and the constant hum of human activity. This is the place that nature marked out as the world's richest source of gold.

But its interest lies not only in the bountifulness of nature and the plenitude of the gold and the minerals being hauled from its depths, important though they are. Interest focuses also on other facets of its long and varied history, on matters which, if not as ancient as the deep-lying gold itself, stretch back through long millennia to reveal the very beginnings of man himself.

And just as the technology and the skills of modern men yield the mineral treasure, so are we now learning more and more about other developments as scientists piece together the historical background and evaluate the importance of other features of the region. The area is rich in history; not only in the saga of recent exploration, settlements and conflicts, but also in fields of vital human interest which, of everlasting significance, stretch back far through the mists of time. It seems as if this area was the cradle of mankind itself.

There are positive indications that this region of perennial water was one of those in Africa where man evolved from the apes which roamed its broad plains and gentle ridges hundreds of thousands of years ago.

Today's Western Deeps mine lies within a broad triangle formed by three significant archaeological sites: the famous Sterkfontein caves, Three Rivers where the Klip River meets the Vaal, and, thirdly, Boskop on the Mooi River near Potchefstroom where a valuable fossilised

Old stables of the Zeederberg Coaching Company, situated near Number 1 Shaft at Western Deep Levels mine.

hominoid skull — of unidentifiable age — was discovered as early as 1913. Those sites have provided archaeological treasures of infinite value in assessing the origins of man, and the triangle within those three apices must be assumed to be a part of the scene where man began. The archaeological sites at the three points, especially Sterkfontein, have provided considerable evidence of early human evolution, data as valuable to scientists studying the history of man as the deep-laid gold of the area is now to the geologist, to the miners — and to the world's bankers.

Professor Revil Mason of the University of the Witwatersrand, writing of the prehistory of the Transvaal, says one "might wish to know why the archaeology of the Transvaal is worth looking at, and whether it has such spectacles to offer as Teotihuacan or Knossos. Although there is nothing as opulent as these, the Transvaal can offer quieter treasures that lie at the roots of human thinking. Here are no golden helmets, no marble gods, no sumptuous temples. All that came later, after the hard work of becoming human had been done, and it is of those earliest upward steps that the Transvaal can boast a record. Here we may claim to add to the general understanding of man's position on earth, his habits and his interests, his mind and his muscles. The Transvaal has a special advantage in this, for it remained undisturbed by civilisation until very recently: not long before Bach wrote his fugues and Spinoza composed his Ethics, Iron Age people broke the first farm lands and displaced the last Stone Age hunters south of the Limpopo. . ."

Professor Raymond Dart's discovery in 1923 at Taung in the northern Cape of the skull of the first australopithecine ("southern ape") — the first in the line of humans — focused attention on the beginnings of man on the African continent. His claims, however, were ahead of their time and his theories were hotly disputed by many internationally prominent scientists. But the discovery by Robert Broom, in August 1936, of the remains of a similar skull in a limestone deposit in a cave at Sterkfontein in the Transvaal, less than 50 kilometres from the present Western Deep Levels mine, confirmed Dart's claims and set archaeologists rethinking and re-examining the prehistory of southern Africa.

Broom's discovery was a fortunate one. So easily could that vitally important skull have been destroyed and lost for ever. The Sterkfontein caves themselves had been discovered by Europeans in the closing years of the nineteenth century and that limestone deposit had already been quarried commercially for some forty years. The precious fossil could easily have been shattered by the blasting. It was only when the caves were visited by the infant Geological Society of South Africa in Johannesburg that their enormous value was appreciated, and the Society persuaded the owners to take steps to prevent their wanton destruction.

At the time, those first two fossils of what were popularly termed the "missing links" between ape and man were the oldest hominoid remains to be uncovered. The world had to wait more than two decades before further evidence came to light. It was not until 1959 that the well-known archaeologist-team of Dr Louis Leakey and his wife, Mary, discovered a similar fossil at Olduvai in Tanganyika, showing that australopithecines may also have evolved elsewhere and probably had roamed widely across much of Africa. Other discoveries of similar, time-related human fossils in Europe and Asia have since followed, while the limestone cave at Sterkfontein and the archaeologically-rich valley at Olduvai have yielded other valuable australopithecine fossils which have contributed to man's knowledge of his past and of the Acheulean culture of which these ancient ancestors were a living part.

In stressing the significance and importance of the Sterkfontein caves, Professor Mason has pointed out that the continuing excavations there have yielded "australopithecines more valuable than all the treasures of the Mediterranean world."

For those early men the plains of the south-western Transvaal around the present Western Deeps mine provided an hospitable habitat. There they left their bones and artefacts in and close to the Sterkfontein cave and at Kromdraai and Swartkrans near at hand, as well as along the banks of the Vaal, the Klip and the Mooi Rivers which not only gave them water for their own sustenance, but which also encouraged plentiful game for man to hunt. Furthermore, periodic flooding of those plains over thousands of years had provided an infinity of water-smoothed pebbles of sizes and of hardness convenient for the shaping of early man's primitive tools and weapons, which people of the Stone Ages there left behind them in abundance — more, indeed, than almost anywhere else in the world.

But not only did Stone Age man leave behind him an abundance of weapons and artefacts, he also bequeathed to future ages magnificent engravings on the rocks. Professor Mason says that "the best Transvaal engravings are amongst the finest art created in prehistoric times . . ."

Men of the Later Stone Age, working in these south-western grasslands of the Transvaal, used flat slabs of rock from ancient diabase dykes, soft enough for their art, yet hard enough to endure for posterity. They engraved in two distinct styles — in different areas — by using a pecking technique or by cutting a continuous outline.

The increasing sophistication of the stone tools of this era has enabled archaeologists to plot the phases of the Stone Age and to trace the development of man himself through hundreds of thousands of years, through those divisions of a continuous series of Stone Ages — Early, Middle and Later — by which his cultural advances have become known and from which his intellectual development has been inferred. And then from stone to metal — iron: progress marked by relics and remains that litter the advancing millennia to the point when, relatively recently, many of the conundrums which teased the minds of archaeologists became riddles for historians.

From the far north into southern Africa, more than a thousand years ago, there came new peoples, bringing with them domesticated animals and different ways of living. They outnumbered by far the scattered Stone Age inhabitants whom they encountered. Settled in established communities, they used the crops which they planted in fields carved from the bush, while the plentiful herds of game dispersed before man's progressive encroachment.

With them these invaders brought a knowledge of fire which enabled them to smelt ores, plentiful in the region, and from the metal they made tools and weapons as well as their simple ornaments.

That invasion brought to a sudden end the Stone Age in southern Africa, and gradually the Iron Age took over. For a while the two cultures co-existed, but inevitably the superior ousted the more primitive. Inexorably the invaders spread across the Transvaal and in due course moved further south, their descendants seen by the early white explorers who ventured into the Transvaal in the middle — and more peaceful — decades of the last century.

Those Iron Age settlements were not to last, however.

The upheaval in the coastal lowlands to the east, brought about by Shaka's violence, had the devastating effect of a tidal wave throughout the land, with consequences that continue to this day. Mzilikazi's invading sweep through the Magaliesberg area decimated the inhabitants and put an end to the Iron Age there.

From the numerous remains scattered throughout the Transvaal, it is presumed that the population increased from mere hundreds at the peak of the Stone Age to hundreds of thousands in the Iron Age. Professor Mason considers this southern region "may prove to be one of the most densely populated areas in Iron Age times." Remains of stone-walled villages are to be traced throughout the area, a notable one built on a low hill at Elandsfontein, near Fochville, close to Western Deeps. For those who care to look, the ruins of many of these old settlements are still to be found, decaying in the bush on the ridges and obscured in the grassy veld in the vicinity of the Western Deep Levels mine. The caves of the Gatsrand have yielded rusted weapons and other iron relics of that past era.

Mzilikazi's clearing of potentially hostile enemies in the south-western Transvaal in the course of his own flight northward made progress easier for the white pioneers from the Cape. Their coming, shortly afterwards, brought a new culture and a fresh dimension to the development of the sparsely populated area across the Vaal River towards the middle of the nineteenth century. It was a development which, within fifty years, was given tumultuous impetus by the discovery of the Witwatersrand gold fields.

The general thrust of the Trek was towards Natal and the Indian Ocean coast, and those pioneers thus tended to steer courses closer to the Drakensberg mountains and through the eastern regions of the Transvaal. Some groups ventured further to the west, however. Among these was the trek led by Andries Hendrik Potgieter. Crossing the Vaal further downstream than the others, he came upon the grassy regions of the Magaliesberg with low dolomitic ridges across plains pitted with natural reservoirs of abundant clear water.

A part of the region came to be called the Gatsrand, in that felicitous way which the Afrikaans language often expressively combines two or more words succinctly to enhance descriptions. Maps today specifically mark the Gatsrand between the Western Deep Levels mine and Carletonville. The characteristic underground caverns were the product of a process which began three million years ago when water in the basic dolomite-limestone began to eat into the undulating rocky earth, leading to the descriptive name given by the early pioneers, "a ridge-with-holes".

There Potgieter decided to settle. Nature was bountiful, water was plentiful, and the grasslands offered substantial grazing for the trekkers' animals. Moreover, the region was further away from the British than either the Free State or Natal.

It was not without threat, however. Mzilikazi, resenting the unauthorised intrusion of the Voortrekkers into this portion of his territory, had earlier murdered a small group of trekkers while Potgieter with some of his men was exploring further north. Punitive action led to the defeat of those Matabele. This, and a drawn-out series of skirmishes, persuaded the defeated Mzilikazi to withdraw northwards, abandoning the area to Potgieter and his handful of pioneers.

The base where Potgieter chose to settle on the gentle Mooi River became known as Potchefstroom, whimsical shorthand that cumbersomely translates into English as the "Stream where Potgieter is chief". In 1840 this became its official name. Of the five settlements then north of the Vaal, Potchefstroom became the first capital of the territory, though it eventually lost its status to Pretoria. Potchefstroom nevertheless had its moments of historical glory.

The town lies some 60 kilometres to the west of Western Deeps and the border of the present magisterial district of Potchefstroom cuts through the mine's property. The northern sector, where Shafts 2 and 3 are located, lies within the magisterial district of Oberholzer. The southern portion, with Number 1 Shaft across the Westwitsrand, falls within the district of Potchefstroom, however.

On the farm Buffelsdoorn, "buffalo thorn", property now encompassed by Anglo American's Elandsrand mine, adjacent to Western Deep Levels, stands an old homestead. For some while this was thought to be the first house built in the Transvaal, but the claim has recently been disproved. The remains of its original owner, a pioneer named Harmse, are buried under a thorn tree where he was killed by a wild buffalo. It was this tragedy which gave the property its name. The grave is still to be seen, now cared for by the mine.

Events were soon to swell the population of the Transvaal, accelerating the development of this quiet, rural area. Of these, the sudden discovery of gold was the most momentous, bringing people from all corners of the world and disturbing the pastoral tranquillity of the Boers.

Lying between the prosperous diamond mining town of Kimberley and the burgeoning wealth of the gold strikes, first in the eastern Transvaal and then the Witwatersrand in 1886, the western Transvaal soon found itself on a busy route from the rail terminus at Kimberley to the new mining areas. For many years that main road passed along the valley in which the Number 1 Shaft of Western Deeps is now situated, and the place echoed with the sound of ringing hooves, of ox-wagons and the creak of laden carriages. Still to be seen on the mine's property are the rusted and decaying remains of the tin-roofed, mud-brick coaching stables, crumbling souvenirs of a vanished past. The rumble of speeding automobiles on the Johannesburg-Potchefstroom highway along the valley's edge now drowns any lingering echo of that romantic horse-drawn age.

Change came swiftly. Four months after George Walker's discovery of outcrop gold on the farm Langlaagte one Sunday morning in February 1886, the Red Star coaching company was reported to be carrying 12 000 pounds of mail a month from the rail-head at Kimberley to Pretoria, travelling via the western Transvaal and Johannesburg. The discoveries on the Witwatersrand quickened exploration which soon yielded finds of gold also in the Potchefstroom area.

A register of the Witwatersrand Gold Fields, published by C.S. Goldman in 1892, lists details of several mines in the western Transvaal. Among them, for example, was Elandslaagte, with property "12 miles south-west of the town of Klerksdorp, which on the 28th February 1891 showed a loss of pounds 7 675 4s 2d." In the Klerksdorp area, too, there was the Eleazer mine. Goldman's lists record that "The financial affairs of this Co. are in a very unsatisfactory state. Nothing is left of the working capital; no proper accounts were kept to show actual expenditure. The secretaries have left the country, and matters remain unexplained." The Emma G.M. Co., Ltd, working "about 17 miles west of Johannesburg", however, showed a profit of £246 14s 2d in its balance sheet of 31 March 1891.

It is interesting to note that the Potchefstroom Exploration G.M. and Estate Co., at that same time with profits of almost £36 000, had properties located "all in district of Potchefstroom, extending in a north-east and south-west direction for about 25 miles; from Johannesburg about 60 miles, from Klerksdorp about 40 miles, from Potchefstroom about 15 miles." Listed among the directors of that company were A. Dunkelsbuhler and B. Oppenheimer, the latter being the eldest brother of Sir Ernest.

But while gold was indeed mined in the western Transvaal in the closing years of the century, the development of the present deep level mines of the West Wits Line had to wait several decades before prospecting and mining techniques had evolved sufficiently to locate the rich reefs, deep underground, and make their exploitation possible.

While the historic significnce of the region undoubtedly continues to lie pre-eminently in its earliest contributions to the study of man, later history has not neglected that region. The events of the Jameson Raid and the Boer War washed like a tide over the area. Facts about the former are well documented, opinions fuelling the controversy. Doornkop, where his party was routed and Jameson surrendered, lies north of the Gatsrand, not far from today's Western Deeps Levels mine.

It was from here, too, during the Boer War — "on the very ground where the Jameson raiders had surrendered four years before" — that war correspondent Winston Churchill "bicycled straight down the main road" into beleaguered Johannesburg.

Prominent on a koppie beside the present busy highway, overlooking the south region of Western Deeps, is a visible reminder of that history. The tall concrete monument to Danie Theron, Boer hero of the war, keeps a stern watch over the valley there. A stone wall encircling the small plateau of the koppie is a relic of an old, long abandoned black settlement.

The slender plinth, capped by a symbolic flame in copper, marks the spot where this hero fell in a skirmish with a detachment of British troops towards the end of the Boer War. The engagement was watched by two women from a farm homestead in the valley, now owned by Western Deeps, a stone's throw away from the old stables of the Zeederberg coach-house.

The young Boer patriot, born at Tulbagh in the Cape Colony, first attained prominence — or rather notoriety — when in April 1899 he thrashed W.F. Moneypenny, editor of the Johannesburg *Star*, for failing to apologise for articles which Theron deemed to be scurrilous and insulting to the Boers.

At the outbreak of the Anglo-Boer War he formed a troop of combatant cyclists, the Wielrijders Rapportgangers, and later became the leader of a troop of military scouts. His courage and reconnoitering skills evoked admiration and his daring brought him wide fame.

In the official history of the regiment, *The Second Battalion Royal Dublin Fusiliers in the South African War*, the authors, Major Romer and A.E. Mainwaring, write of the engagement on that koppie which was Danie Theron's courageous last stand. "... After moving some two or three miles," they write, "our further way was blocked by mauser-fire from a very ominous black-looking kopje which stretched down into the valley from the high ground on our left. The guns came into action against the hill at a range of about two thousand yards, and it seemed as if a golden crested wren could not have escaped if it had been unlucky enough to be there. The shrapnel kept up an almost incessant hail ... while every now and then the deeper note of the 4.7 was followed by a huge cloud of dust and yellowish vapour thrown up, and off, by the explosion of the lyddite in the huge projectile. ..."

The Boer's version of that engagement is told by J.H. Breytenbach in his book on Kommandant Danie Theron.

"*... In hierdie vreeslike kartets- en liddiethael, waaruit volgens Britse mening, geen tinktinkie kon ontkom nie, het Danie alleen gestaan. Tussen die geblaf van die kanonne, en doe gefluit en gekraak van kartetse en bomme, het sy Mauser aangehou om die kakies wys te maak dat daardie rantjie vol Boere is wat nie bang is nie — aangehou totdat 'n Engelse bom hier vlak naas hom teen 'n klip vasslaan en dit aan stukke ruk. Toe eensklaps het sy Mauser geswyg. Danie Theron was daar nie meer nie. Een van die klipskerwe het hom aan die een kant aan die skouer, die bors en die kakebeen getref, en hom teen die grond neergewerp — vermoedelik onmiddelik gedood. ...*"

That monument crystallises a brief moment in an infinite history of the region that is recorded in a variety of ways from ages before man himself was shaped: history to which, in its own way, Western Deep Levels is now contributing.

CHAPTER FOUR

How deep is "Deep Level"?

Though the controversy is now overlain with the dust of time, who first really discovered the gold of the Witwatersrand remains yet a matter of some — if dwindling — conjecture. Such was the extent of the prolonged debate some years ago, however, that in 1938 the South African government set up a commission to enquire into the historic claims of various prospectors. The ensuing Report, not without certain contradictions, discusses the related claims, but provides no categoric answer; rather, it evades commitment, leaving the question almost as open as it ever was.

The Report states, for example, that while Pieter Jacob Marais was "the first person to find gold near the Witwatersrand", the Struben brothers "by their extensive and persistent prospecting and mining activities on the Witwatersrand, both north and south of the Main Reef, . . . attracted so much attention to this area, that the subsequent discovery of the Main Reef series became inevitable." But the Struben brothers were not the only prospectors to have contributed to the knowledge of gold deposits in the vicinity of the Witwatersrand. Others found alluvial gold between Pretoria and Potchefstroom and, also, south of the ridge, in areas reaching to Heidelberg.

The first discovery of the Main Reef, the Report continues, was made on the farm Langlaagte, "probably in the last ten days of March, 1886." It was made by George Walker, in conjunction with George Harrison, whose own version of the momentous event indicates that the discovery was in fact made in February — not in March — of that year. Walker, one Sunday morning on his way to visit his friend working as a builder on the farm Langlaagte, stumbled over an outcrop of rock in the long grass, "recognised it as conglomerate, panned it, and found it to be rich in gold." That Harrison did not regard the find as at all promising, however, may be assumed from the fact that he sold his claim shortly afterwards and headed for the goldfields at Barberton, never to be seen again. The story of their find — of ore richer by far than ever could have been imagined — has been recounted often enough.

"This discovery was accidental", the Commission goes on to record, though "the presence of gold in the Transvaal seems to have been suspected very soon after its first occupation by the Trekkers." Perhaps this is not to be wondered at, since those pioneers had moved into regions which for centuries had been marked on old maps as part of the fabled kingdom of Monomotapa, "where gold had been found from time immemorial."

It is difficult to pinpoint the beginnings of those nineteenth-century searches for gold in the Transvaal. Initially they seem to have been haphazard rather than deliberate, made by explorers and hunters, rather than prospectors, often stumbling casually upon alluvial deposits. Sir John Barrow, an English explorer who travelled north of the Vaal River, as early as 1806 reported the country to be rich in minerals. On a map of his explorations he noted that mountains, roughly where the Witwatersrand is, contained gold. Another Englishman, John Henry Davis, was among the first whites to discover gold across the Vaal River. Having studied mineralogy, he was one of the few able authoritatively to identify as gold the substance he had found in 1852 near what is now Krugersdorp, about 50 kilometres from where Western Deeps is today. The find was hushed up, however, for when he reported his discovery to President Pretorius he was financially compensated for the gold he had recovered, sworn to secrecy and ordered out of the country by the President, terrified that news of the find would bring a rush of foreigners who would both outnumber and swamp his people and disturb their rustic tranquillity.

The first recorded accounts of whites actually prospecting for gold in the Transvaal would seem to date from Pieter Marais' appointment by the Volksraad in Potchefstroom as an official prospector. The terms of his appointment in 1853 similarly demanded absolute secrecy, any breach of which that might cause "the independence of the Republic to be disturbed in any way . . . shall be punished with the penalty of death and no extenuating circumstances will be taken into consideration."

Marais was born at the Cape. An unsuccessful attempt at gold-mining in California in 1850 was followed by another — scarcely more fortunate — in Australia, and he soon returned to South Africa. In September 1853, surely with gold as his objective, he headed northwards to Potchefstroom, where he immediately made himself known to the authorities by contacting the landdrost.

History seems not to have recorded what was then discussed, but, given Marais' background, it may safely be assumed that the mining experience he had acquired in California and Australia did not go unmentioned. Not only did the Volksraad give him permission to search for gold, but, later, also promised to reward him for any substantial discoveries he might make.

His first weeks were frantic with searches in the immediate vicinity, but without success. Gradually extending his prospecting beyond the western Transvaal, he reached eastward to the Jukskei River, between what are now Johannesburg and Pretoria. There luck smiled upon him, albeit faintly, so that within five weeks of his arrival in the Republic he was able to note in his working diary that he had panned gold in the Jukskei River.

In the report of the 1938 Commission, Marais is thus credited with the discovery of the first gold to be found near the Witwatersrand, insignificant though the results were.

The finds encouraged him. But luck's smile was destined to become a cruel leer. Both in California and in Australia, Marais had panned for alluvial gold in rivers and streams, as circumstances demanded. Heir to that experience, he did the same in the Transvaal. The Transvaal deposits were different, however. Here the gold lay not in nuggets and grains in the rivers. It seemed not to occur to him to look elsewhere — among the rocky outcrops of the highveld, for example — or to seek the source of the paltry specks of gold he had found in the streams. Unthinking and hide-bound by experience, he thus missed the main outcrops of the gold reef on the ridges.

From the Jukskei he turned again to the streams of the western Transvaal. Again his labours came to naught as swirl after swirl of his pan revealed no tell-tale golden glint in the hot summer's sunshine.

According to the report of the Commission in 1938, the Volksraad's promised reward for payable gold found by Marais was fixed at £5 000 — for those times a vast sum indeed, so great, in fact, that some historians consider £500 a more likely figure. But however large the promised reward, given the prevailing bankruptcy of the Transvaal Republic, it would indicate a desperate keenness to find a source of viable wealth to establish the solvency and prosperity of the Republic. The fears of likely political and social consequences of gold discoveries in the Transvaal were outweighed by the impoverished government's desperate need for resources. Hopes were diluted by fears.

Looking north across Number 2 Shaft complex at Western Deep Levels mine.

Marais, the internationally experienced miner upon whom hopes had been pinned, provided no solution, however, and the dark clouds of poverty continued to threaten the Republic, while the ambivalent fears of political submersion, never far in the background, did not dissipate either.

Eventually, his quest unfulfilled, Marais was dismissed by a dissatisfied Volksraad and left the Transvaal a few days short of two years since he had first crossed the Vaal. He was the first official gold prospector in the Transvaal. There is no record of any other.

His travels between Potchefstroom and the Jukskei River must have taken him often over the ground where the Western Deep Levels mine is now situated and frequently he must have traversed "the long watershed known as the Witwatersrand".

Stimulated by sporadic discoveries of gold in various parts of the Transvaal, hope lived on. Finds were frequent enough to lead the Volksraad, still fearful of foreign intervention, to pass what has become known as the First Gold Law of the Transvaal: No 1 of 1871. Its provisions set a precedent that was repeated in the Transvaal and, later, were invested in South Africa's mining law.

Article 1 reserves to the State the right to mine all precious minerals. The State claims neither the gold nor the precious stones themselves, but rigidly controls the right to mine them.

This entrenched right influences mining taxation in South Africa. Since title to all precious minerals is invested in the State, payments are made to the State as the lessors.

The discovery of gold at Eersteling in 1871 was followed by others in the Transvaal. Traces were found in the western Transvaal. On the farm Weltevreden, near what is now Fochville, shafts were dug on an outcrop of conglomerate in about 1881, but the efforts proved fruitless. Finds near Potchefstroom were followed by more not far from Rustenburg. Over in the eastern Transvaal, greater success at Barberton in 1884 increased expectations and, from all over the world, brought miners whose experience became a boon for the Witwatersrand discoveries two years later. Their success at Barberton was short-lived, but their later value to an industry about to explode into life was immense.

Eersteling and Barberton, not least, had provided new lessons in the extraction of gold. Gold there was not alluvial, to be gently panned from rivers and streams. It had to be won from hard quartzite. The introduction of miners with technical skills unknown to the simple alluvial prospector was to prove invaluable. Hammers and pans became obsolescent, and soon obsolete as machines took over.

Miners with experience came from Barberton and capital was plentiful in Kimberley. The two met in Johannesburg. That meeting of skills and money was a fortunate coincidence for the Witwatersrand. Investment flowed into the Transvaal as diamond companies and financiers hastened to the Witwatersrand seeking to spread their surplus capital into fresh — and hopefully lucrative — ventures.

But though Kimberley's money played a vital part in the establishment of companies to work the deposits of the Witwatersrand, the skills which the diamond industry had to offer were not suitable for gold mining. The situation was further complicated by the fact that there was no technical precedent for mining the reefs of the Witwatersrand. A report of the Chamber of Mines in 1902 pointed out that "up to the time of the discovery of the Witwatersrand deposits, no auriferous conglomerates had been worked on a large industrial basis in any part of the world." New methods had to be pioneered.

In the short months between the epic discovery by Walker and Harrison in February 1886 and the formal proclaiming of the Witwatersrand goldfields on 4 October, numbers of prominent financiers from Kimberley sought to establish prospecting and mining companies to exploit the new discoveries. Among them were Cecil Rhodes, Barney Barnato, Alfred Beit and J.B. Robinson.

The importance of readily available capital for the diggings cannot be overemphasised. The Standard Bank's Manager in Kimberley, reporting to his Head Office on the situation in 1886, said: "Local commercial men are endeavouring to secure a portion of the Transvaal trade ...

Kimberley will we think be an important place in connection with the development of the new gold fields . . ."

For the Transvaal the discovery of gold came in the nick of time. In his memoirs President Kruger relates how desperately in need of money the Republic then was; he makes it plain, however, that the burghers would have preferred succour in some less ambivalent form.

"The finances were in a sad condition," he said. "The credit with the Standard Bank had become exhausted and they refused to advance more money. . . . The rich gold-fields of the Witwatersrand were discovered and brought about a complete revolution in the financial aspects of the affairs of the Republic. The history of the Republic entered upon a new phase with this discovery. Can we possibly look upon it as fortunate? . . ." He goes on to quote the words uttered by General Joubert . . . "Instead of rejoicing, you would do better to weep; for this gold will cause our country to be soaked in blood."

The extent of that "complete revolution" may be gauged from a Standard Bank report written early in 1887: "Twelve months ago the Republic's finances were in a very impoverished condition, whereas now the Government's credit balance with our Pretoria Branch is by latest advices 60 000 pounds . . ." In no time the Transvaal had changed from being "the most retrogressive of South African states" to the "mainspring of such activity as at present prevails in this country . . ."

By today's standards early mining methods on the Witwatersrand were nothing if not primitive. Black labourers merely dug open trenches to exploit the outcrops that were visible and shovelled the gold-bearing ore to the surface. As the trenches went deeper — beyond the feasible limits of shovelling — oxen were used to drag the ore to the surface.

Conditioned perhaps by the formation of the gold reef at Barberton where mining proved the strata "to be almost vertical or even overturned", the early prospectors on the Witwatersrand appeared to think the reef there had a similar formation. Diggers had yet to comprehend the eccentricities of this new reef and, unaware of its formation, failed to understand how it dipped into the earth, angling away from the surface and eventually below reach.

At the outset, work was thus concentrated on the visible outcrops within the proclaimed area, and the limitations became more and more obvious as the surface gold petered out and shallow workings were abandoned. At hand there happened to be an experienced American mining engineer, Joseph Curtis, who, understanding the anomalies of the Witwatersrand reef, confidentially advised his employers, Eckstein and Company, without fanfare to purchase ground to the south of the outcrops of the officially proclaimed area. This stretched several miles, straddling the east-west axis of the main reef.

Companies privy to advice from Curtis and other American engineers soon took possession of the ground further south, so that for some 15 miles or so along the reef, the activities of the outcrop companies and of the "deep-level" operators marched in parallel across the landscape.

Curtis, "one of the most eminent of mining geologists" of his day, had been responsible for the geological survey of the United States. Professional curiosity brought him to the Transvaal in March 1887 and at once he began an intensive study of the Witwatersrand gold reef, a study which quickly changed the concepts of mining in South Africa and inaugurated the system of underground mining as we know it today. He was the first of several prominent American mining engineers who, in one way or another, made valuable contributions to the technical development of South Africa's gold mining.

It was they who pioneered an understanding of the geology of the Witwatersrand, appreciating that the outcrops at Langlaagte were merely part of a far larger reef which, dipping below the surface, extended well beyond the proclaimed area. At that time none was able to gauge the full extent of the dip, however; nor, of course, could geologists then foresee how the deeper gold-bearing reef might be overlaid with other strata.

Curtis's predictions were proved early in 1890 when test drilling, begun four months before, intersected both the South and Main Reefs at depths of 517 and 581 feet respectively (approx. 160 and 180 metres). Soon afterwards extensions of the reef further down dip were confirmed by evidence produced from other boreholes.

This removed any lingering doubts that the winning of gold was still an adventure for the itinerant prospector, at any rate on the Witwatersrand. The needs of the new system and the magnitude of its costs now made profitable gold mining increasingly the prerogative of big business as activity left the surface and, like a mole, began to move underground.

"Deep level" became a popular description of the new methods that suddenly were necessary. But the description then by no means matched the concepts of today. At that time, the term merely signified any mining carried on below ground level. No qualifying limits were defined.

Not everyone, however, accepted the theories and recommendations of the American engineers. J.B. Robinson, with a number of other mining financiers in search of new investments, had moved to supplement his diamond operations at Kimberley with the profitable gold of Johannesburg. He publicly denounced deep-level mining as a fraud and offered £1 000 to anyone who could prove that a shaft sunk 100 yards from the proclaimed area would strike gold.

But sceptical — or perhaps cunning — though he clearly showed himself to be about the future of deep mining, Robinson was soon forced to change his mind as the validity of the prevailing theories became increasingly evident. Notwithstanding his professed scepticism, he did not baulk at mining well below the surface on the properties he had acquired. It is thus not without a certain irony that the report of the official enquiry of 1938 into the discovery of gold on the Witwatersrand states that evidence "shows beyond dispute that J.B. Robinson can claim the credit of being the first to inaugurate deeper mining on a large scale on the Witwatersrand." That involved the digging of incline shafts to "a depth varying from 65 to 75 feet", and one incline shaft extending to a depth of 115 feet.

Eckstein himself, despite the expert advice to which he was privy, had also been sceptical about the future of deep underground mining. If the theories were right, he thought, shafts might have to reach down to between 800 and 1 600 feet before striking the ore-bearing reef. Furthermore, he estimated, such shafts would cost at least £50 a foot to drill and would probably take as long as two or three years to complete. "This tells its own tale," he had cautioned friends in London. With technical and financial problems of such magnitude he felt such "enterprises . . . cannot possibly live."

Notwithstanding Eckstein's earlier personal caution, however, his firm was among the first to launch into deep-level mining and to establish the company, Rand Mines, which controlled them.

Subsequent history has shown the accuracy also of the prognostications of Hamilton Smith, another American mining engineer who visited the Witwatersrand in 1892. With considerable accuracy, it now turns out, he gave a remarkably clear picture of the geological formation of the Wits reef and basin, and predicted the future discovery of the southern rim at some point south of the Vaal River.

Furthermore, he foresaw the levelling out of the dipping reef with continuing payable gold values, and considered that neither the obstacles of heat nor underground water would inhibit mining before a vertical depth of 3 000 feet (approximately 930 metres).

Though perhaps puny by today's standards, and certainly well short of the present achievements and future plans of Western Deep Levels, the early concepts of mining at depth continued to expand, and we read of the "Deep-deep Agreement" of 1894, in terms of which three established mining companies committed themselves to the joint exploitation of only deep-level mines (that is, underground mining) on the Witwatersrand gold-fields.

In August 1894, The Standard Bank reported to London that "Deep Level Mining is being prosecuted with great vigour . . ." and the following year added: "Greater attention than ever is being paid to Deep Level Mining, and the lives of several of the outcrop Companies, which it was feared were of comparatively short duration, have been lengthened by dip claims . . . there can now be little, if any, doubt as to the permanence of these fields, and the stability of the industry . . . for many years to come . . .".

In 1897 a circular issued by Consolidated Gold Fields explained the concept of deep-level mining in relation to outcrop mining. The explanation is an interesting one. Deep-level mining was there defined as "the mining of the reefs below the points where they are owned by the outcrop Companies". And the circular went on to point out that "the term 'deep level' is a relative one."

That pamphlet continued: "Our present deep level Companies contemplate striking the reef in all instances within a vertical depth of 3 000 feet, some of them indeed at considerably lesser depths. These depths are shallow compared to the depths at which mining is successfully carried on in America and Europe.

"From an economic point of view, the difference between deep level mining and the mining of an outcrop property is that in deep level mining vertical shafts must be sunk to strike the reef after which the methods of mining are identical; while in the case of the outcrop mine the sinking of this vertical shaft is unnecessary."

As techniques developed and as knowledge of the geology of the reef expanded, shafts sank deeper and deeper, while, further afield in a widening area, options on properties were purchased by companies keenly looking to the future.

Cost went up as the mining went down. Not only were costs inflated by the necessarily increased physical operations, but they were also increased by problems in the changing compositions of the host ores. As the oxidised zone near the surface gave way to the sulphide zone, gold recoveries began to fall from 75 to 60 per cent. As it became clear that the old mercury amalgam process would not be able to treat the local sulphide ores efficiently, panic took hold of the Stock Exchange and the value of Witwatersrand gold-mining shares fell dramatically.

Processes were devised to meet the problems set by these Witwatersrand ores. The use of chemicals enhanced the efficiency of recovery, but about a quarter of the gold still remained tantalisingly out of reach.

A most valuable breakthrough for the South African gold-mining industry came with the introduction of the MacArthur-Forrest cyanide zinc recovery process in the late 1880s.

A tremendous boon, this development led to an overall recovery rate of about 90 per cent and, as the years passed, with the introduction of finer grinding and the "all slimes process", efficiency rose steadily to 96 per cent. The value of the contribution of this process to the gold mining industry cannot be overestimated.

The Transvaal Chamber of Mines — first formed in 1887 and reconstituted in 1889 — records that in 1892, apart from the operations on the Witwatersrand, there were more than a dozen goldfields being worked in the Transvaal and that these areas in that year produced some 114 527 ounces of fine gold. At the time of the discovery of the main Witwatersrand reef in 1886, the Transvaal produced a mere 0,16 per cent of the world's gold. Growth, slow at first, was phenomenal, however. By 1898 the Transvaal had become the world's biggest producer, overtaking Australia and America, contributing 27,55 per cent of the world's supply. The wavy line of the main reef had been traced and by the end of the century mines were operating along its known length from Randfontein to Modderfontein, 30 per cent of them being described as "deep level" in the concept of those times.

General Joubert's words proved prophetic. Gold mining was indeed interrupted by bloodshed and weeping, brought about, firstly, by the Jameson Raid and not long afterwards by the Anglo-Boer War. Gold was central to the issues at stake.

Professor Geoffrey Blainey of Melbourne University, in an interesting sally into South African history, considers that the circumstances in which deep-level mining found itself under President Kruger in the 1890s sparked off the Jameson Raid.

"The Jameson Raid," he has suggested, "was essentially the revolt of the two big companies that were heirs to the treasures and problems of the deep levels."

He points out that profits from Witwatersrand gold mining in the early 1890s came almost entirely from the outcrop mines, while the deep-level mines — "gluttonous for capital in comparison with the outcrop companies" — had to bear not only the high costs of shaft sinking but also the burden of delayed results. On top of that they suffered also from the burdens of Kruger's taxation policies. Requiring far greater working amounts of dynamite, for example, than the outcrop operations, the deep-level mines suffered unduly at the hands of the "dynamite monopoly" whose charges were inflated by Kruger's system of taxation. Deep-level mining was further penalised — compared with outcrop operations — by the labour policies of the Transvaal government.

These and other complaints of bias against the deep-level mining companies have prompted Professor Blainey to suggest that they were at the heart of the revolt against the Kruger government which led to the Jameson Raid and, later, to a lingering antipathy which exploded into the Boer War.

During the war, it was common cause among both the Boers and the British authorities that destruction of the mines would benefit neither side in the long run and special measures taken for their protection during hostilities were rigidly observed.

The war over, attention turned again to the feasibility of mining in the area west of Johannesburg and to finding the main reef which dipped out of reach into the depths of the encompassing dolomite. In 1902, "after years of careful and diligent prospecting work and the expenditure of very large sums of money," as its Chairman explained, Western Rand Estates was formed to exploit the area south-west of Randfontein.

The western Transvaal had not been overlooked in the initial rush for gold and a number of options on properties had been taken out in 1887. Unexploited, however, the options had lapsed. But in the closing decade of the century, mines had been successfully established in both the Potchefstroom and Klerksdorp areas. In 1898 Pullinger began the first exploratory borehole in the district east of Potchefstroom, on the farm Middelvlei, one of those on which Consolidated Gold Fields had earlier taken options, which had lapsed. With the outbreak of the Boer War, Pullinger's operations were suspended and it was not until the conclusion of hostilities that drilling was resumed, by Western Rand Estates which had absorbed Pullinger's holdings. Shares were offered to several mining houses but none wished to participate, despite promising assay reports. Co-operating with Western Rand Estates, Pullinger persisted and in 1910 began sinking a shaft on Gemsbokfontein, now part of the property of the Venterspost mine, not far to the north-east of Western Deeps.

Work had scarcely got under way, however, when, at a depth of slightly less than 30 metres, the shaft suddenly encountered water in such quantities that operations had to be suspended. When pumping failed to cope with the gush of five million gallons a day, exploration was abandoned. It became clear that the drilling site had been selected unduly close to the Wonderfontein Spruit. Various attempts in the following years to resuscitate the venture proved in vain. Nor was the necessary capital forthcoming. But the consequences of the setback went further than that: for a number of years the calamity had a negative effect on further prospecting along the West Wits Line, until indeed the arrival in South Africa of the cementation process, a Belgian development, by which water-bearing fissures could be plugged by cement mixtures.

The introduction of this process was to have a tremendous impact upon the country's mining industry. Without it, several operations would have been rendered more costly and more difficult — if not impossible — particularly in the rich areas of the West Wits Line where the dolomite is riddled with fissures and vast underground caverns that brim with water trapped through long ages. J.S. Crawhall, former Managing Director of the Cementation Company (Africa), has described the area as having "probably the greatest concentration of water problems in shaft-sinking anywhere in the world," adding that "Shaft-sinking, in most cases, has been a Herculean task" there.

Interestingly, although the cementation process was first introduced to South Africa as early as 1916, several years after the setback of the unsuccessful Pullinger Shaft, and was utilised — notably by East Rand Proprietary Mine (ERPM) — to solve problems of underground water on mines, it first proved its worth in the West Wits area only in 1934. Gold Fields, in the early stages of an enterprise that was to launch a fresh era in South African gold mining, began sinking shafts for the Venterspost mine, the property where Pullinger's efforts had failed against the underground water almost a quarter of a century earlier.

That operation at Venterspost involved the drilling of two shafts, 3,6 kilometres apart and some distance to the east of the old Pullinger shaft and away from the Wonderfontein Spruit. At both shafts water was encountered at just below 30 metres underground and again at 61 metres — and yet again at 1 390 metres. It took a million bags of cement to stanch the flow.

The success of the cementation process set a new pattern for mining in South Africa. It is now standard practice in shaft sinking in any mining area where the presence of underground water is known or suspected.

The process involves the drilling of pilot boreholes, usually of approximately 3,8 centimetres in diameter, to ascertain whether water is present in the vicinity of the proposed shaft or stope. Should the probe

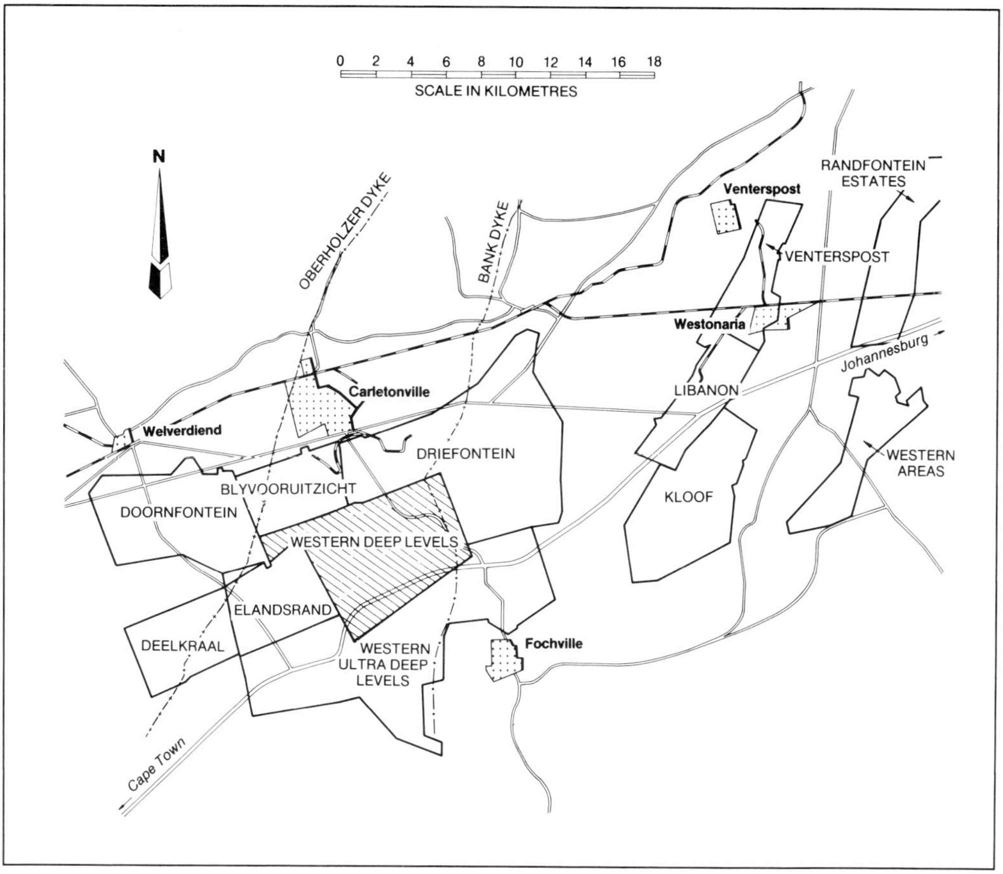

Western Deep Levels Limited in the West Wits Line.

encounter a water-bearing fissure, a stream of water and cement mixture is injected under intense pressure into the cavity. The cement there deposits itself on the walls of the cavity, layer upon layer, until the fissure is sealed strongly enough to contain the forces of the natural pressures.

The pilot holes, surrounding the shaft in its course of sinking, extend — usually 20 metres or so — ahead of the shaft itself. Should a pilot hole encounter a fissure, shaft-sinking is halted and cementation is immediately applied until the fissure is sealed.

The number of pilot holes surrounding a shaft in the course of sinking depends largely upon the size of the shaft and the nature of the ground. On the West Rand twenty are generally accepted as the minimum number for a standard rectangular shaft while a ring of sixteen is the minimum required for a circular shaft.

The 1920s were not happy years for the gold mining industry. A report by the Government Mining Engineer in 1925 listed a number of mines on the East Rand in which substantial capital had been invested but which, in the face of escalating costs, had either ceased producing or, worse, had closed down before reaching the production stage. Without the prospect of fresh mines opening, the end of the Witwatersrand gold bonanza appeared to be looming close.

The report went on to list a number of "the more promising properties" in the area still available for mining and which were "estimated to contain the Main Reef Series at depths less than 7,500 feet" (2 325 metres).

But in the face of the prevailing stringency of the money market in those Depression years and the consequent difficulties in raising capital, either by shares or loans, the outlook for such developments was bleak indeed.

South Africa's gold mining industry looked to some *Deus ex machina* for relief — and, indeed, for its continued livelihood. After all, if the prognostications of the Government Mining Engineer in 1925 were valid and existing mines were shortly to become exhausted, then to all intents and purposes the industry would be defunct by the middle of the century. This depressing opinion was confirmed some five years later. Dr Hans Pirow, who had succeeded Sir Robert Kotze as Government Mining Engineer, predicted that by the middle of the century South Africa's output of gold would have shrunk to less than a quarter of that produced in 1930. For a country that derived half its revenue from gold mining the signs were ominous indeed. No less bleak were the prospects for the nation's workers, about 50 per cent of whom gained their livelihood from the gold mining industry.

In 1931 Britain, jostled by the Depression, left the gold standard. The South African government, obstinately ignoring advice and in a brave show of political and constitutional independence, which many saw as a belated shot from the Boer War, decided not to follow suit. For a while, until the worsening Depression and the political pressures it generated throughout the land forced the government at the end of 1932 also to abandon the gold standard, the country tottered on the brink of economic disaster.

At once the situation changed dramatically. The price of gold rose sharply, by almost 50 per cent. By May of that year, the market value of gold shares on the Johannesburg Stock Exchange had increased from the previous December's figure of £125-million to £300-million.

Throughout the country the economic depression began to lift. The wheels of industry began to turn again and there were fresh hopes for the gold mines: numbers of lower-grade operations would now become profitable.

At the same time, the growing confidence and prosperity increased the feasibility of plans for the development of new mines to offset the dwindling output of older operations, which together would help considerably to lift the industry out of its state of economic stagnation.

Once again in the history of mining on the Witwatersrand, circumstances developed to present man with fresh challenges.

For many years geologists had been puzzled by the enigma of the auriferous main reef, which, south of Randfontein, vanished into the earth beneath thick layers of ancient lava. The results of test boreholes drilled in the 1920s along the expected line of the reef were not sufficiently positive to induce mining houses to proceed with the exploitation of options in the western area. None could say whether the gold-bearing reef came to an end, or whether it merely vanished beyond reach. Risks were high and on the information available were beyond justifiable working limits. Geologists were cautious in their opinions, conservative — even negative — in their prognostications and suggestions.

But if the theatrical machine did not produce a god, suddenly the gods produced a machine! And at the opportune moment!

A positive consequence of the world depression of the 1930s was the arrival in South Africa of a young scientist, Dr Rudolf Krahmann, who, with his wife, had left his native Germany to seek the living he was unable to find in a depressed Europe. This proved to be a windfall for South Africa. He was destined to play a vital part in the development of our mining industry, particularly in the discovery and development of the West Wits Line — and ultimately, therefore, in the establishment of Western Deep Levels.

At the University of Berlin, where his father was head of the faculty of Economic Mining, young Rudolf Krahmann had earned a doctorate in Engineering and Mining Engineering, specialising in economic geology. He had later worked at the university and then, in the employ of the Elbhof Geophysical Prospecting Company, had done research in several parts of the world. This had led him to the development of a new method of prospecting, using an instrument known as a magnetometer.

How he came to apply these theories and methods to mining geology on the Witwatersrand is a story that is no less fascinating for having been often told.

On 11 June 1930, a few months after arriving in the Transvaal, Krahmann and his wife were invited to watch a motor race at Muldersdrift near Krugersdorp. His own words relate best the great moment on that bright winter's afternoon when he was met by the sight of what for him, new to South Africa, must have been an unusual geological formation. "In the sharp glare of the sun my attention was drawn by the glitter of the surrounding shales," he recorded. "This together with the reddish colouring of the soil indicated a high iron content. If so, the glitter must originate from minute magnetic crystals. And so it proved to be; by moving about a barely pea-sized fragment of shale, I could turn the needle of my geologist's compass in any direction I liked. While playing around with this shale fragment, the idea struck me whether it would not be possible to use this ferruginous shale horizon of the Lower Witwatersrand System as a guide to locate the Main Reef of the Upper Witwatersrand System which disappears under young rocks south of Randfontein."

He later went on to say: "My idea — it was the egg of Columbus — was to ignore the direct approach to the gold-bearing reefs and to study the physical characteristics of the accompanying rocks. Of these the magnetic content of certain horizons was the most obvious and appeared the most promising. These iron-rich shales of the Lower Witwatersrand System, occurring at reasonably constant levels below the Main Reef, could then be used, indirectly, as markers to trace the continuation of the gold-bearing horizons of the Upper Witwatersrand System."

He postulated that magnetometer observations would trace the position of the magnetic shales, which in turn would give clues to the whereabouts of the Main Reef known elsewhere to lie above several strata of magnetic shales. It would then remain for boreholes precisely to locate the actual reef and for assayers in the usual way to gauge the wealth of its content.

He put his theory to the well-known geologist, Dr Hans Merensky, who, greatly impressed, at once helped the penurious Krahmann to purchase a magnetometer from Germany. Field tests with the instrument soon convinced Krahmann that his theories were almost certainly correct, in which case the potential value of his discoveries would be enormous.

Who now to turn to? Merensky had an overload of hay on his fork and seemed disinclined to add to it. He directed Krahmann to another German geologist in South Africa, Leopold Reinecke, who suggested that Krahmann should contact Gold Fields. There he met Carleton

Jones, the company's Consulting Engineer, the importance of whose role in the development of the West Wits Line is enshrined in the name of the town, Carletonville, on the north side of the Gatsrand, about eight kilometres away from Western Deep Levels.

Carleton Jones was captivated by the prospect of a possible solution to the riddle of the vanishing Main Reef. It was something which had long puzzled him too. At once he hired Krahmann. Swearing him to secrecy, he sent him out with his magnetometer to survey the area south-west of Randfontein along those valleys as far as the Mooi River, a distance of more than 60 kilometres.

On 28 July, just over six weeks after the rocks at the site of the motor races at Muldersdrift had set him thinking and while he must have been mulling over his theories, Dr Krahmann gave a lecture to the Geological Society of South Africa. He spoke on "Magnetic Investigations as an Aid to economic Geology." Nobody listening to that talk in Johannesburg could have gained an inkling of the momentous thinking then hidden in the speaker's mind, nor could anyone have realised from what he said that the imminent application of those theories would launch South Africa's mining industry into a new era of prosperity.

Dr Krahmann opened with "a few remarks on one of the most important four geophysical methods" of geological surveying, and told his audience that "the preponderance of rather old formations and extensive granite masses in the geology of South Africa" was the reason why torsion balances and seismometers had scarcely been used here. He suggested that "for the geological and mining problems in South Africa, the electric and magnetic methods of investigation were the most suitable." No more than vague generalisations.

Krahmann went on to relate instances of the success of such surveys in other parts of the world — the United States, Germany, Russia and Spain. But not once did he give the slightest hint of what must then certainly have been uppermost in his mind, his thoughts whirling about the future as he spoke of the past. "Concerning the economic geology of South Africa," he said, "I am sorry to have to say that as yet I am not sufficiently well versed in local problems to suggest propositions for magnetic investigation."

What followed his meetings with Reinecke, Merensky and Carleton Jones is now history.

"As a result of information gained from a geological and magnetometric investigation of the area lying to the south-west of Randfontein and extending west to the Mooi River," Gold Fields in 1932 formed

West Witwatersrand Areas to follow up the survey with extensive borehole drilling and to look to the purchase of whatever properties and options it could lay hands upon. The bogey of the underground water in that area now laid to rest by cementation, Gold Fields appeared to be on the crest of a new wave.

Carleton Jones offered other mining houses a share in the development of the area and was disappointed when few showed interest. On the basis of information confidentially made available with the limited offers, the Consulting Engineer of the Anglo American Corporation, Frank Unger, in October 1932 told Sir Ernest Oppenheimer of the "tremendous possibilities" which the area offered and which, "if there is any luck", would lead one to override all objections to participation in mining in that section of the Far West Rand.

Unger admitted that the scheme was "a gamble", though "certainly a very attractive one, and of a type in which a mining house is absolutely justified in risking even a considerable sum of money."

The Anglo American Corporation showed positive interest and bought into the action with a contribution of 15 per cent of the initial share capital of the venture.

Almost from the start of developments on the West Rand Anglo American had shown considerable interest in the area, participating in the exploitation of ground south-west of Randfontein and westwards to the Mooi River. For this purpose West Witwatersrand Areas had been created. In the following years the Corporation increased its shareholding in that company by exercising its rights to a share in the issue of debentures and also by purchasing shares on the open market.

In his report to the shareholders of the Anglo American Corporation at a meeting held on 30 April 1937, the Chairman, Sir Ernest Oppenheimer, announced the formation of the West Rand Investment Trust. He became its chairman. Its functions were to acquire, *inter alia*, interests in the Klerksdorp and West Rand districts and to take over the Corporation's holdings in West Witwatersrand Areas and its share in the Western Reefs Exploration Company.

Sir Ernest spoke frankly, pointing out that little was known about the areas over which the Corporation had now acquired options, but insisted that the venture was not in the "speculative" class, in the sense that "speculative" implied "the engagement in operations that involved the risk of total loss." The consulting engineers both of the Corporation and the New Consolidated Gold Fields had indicated that the proposed projects had "reasonable chances of success".

A year later, at the first annual general meeting of the newly-formed

West Rand Investment Trust, Sir Ernest said that the Board intended putting forward proposals "for a more permanent method of financing the company than the present arrangement of borrowing from the Anglo American Corporation", but, "with present market conditions the time is not opportune to consider such a scheme."

While hostilities interfered severely with Anglo American's plans for the exploitation and development of the new mining areas then becoming available, planning and preparation for the post-war future prudently continued. This foresight enabled the Corporation to minimise delays when hostilities ended, so that it was only in the fifties that the exploitation of the area south-west of Randfontein, along the West Wits Line, really got under way.

As the enormous wealth of the area began to reveal itself, others began taking greater interest in the region and several new companies, notably Blyvooruitzicht Gold Mining Company Limited, were born. In spite of the limitations imposed by hostilities, Venterspost continued in production, while Blyvooruitzicht began producing during the war years. War or no war, the Allies still needed South Africa's gold.

In droves miners from the Transvaal volunteered for service in South Africa's forces, and went to war as soldiers, sailors and airmen to serve, first in Africa and then wherever volunteers were needed. In many theatres of hostilities the skills and expertise of South Africa's miners added to the splendid contribution which Springbok troops made to the Allied war effort.

On the West Wits Line test drilling came to a halt. Incomplete boreholes were sealed, awaiting peace. In the previous sixty years the tides of South African wars had twice washed over the area, and, too, this was the second occasion in a quarter of a century that men from the farms and hamlets of that region had gone forth to defend liberty in faraway lands. In 1939 they were reminded of this by a town in their midst — Fochville — named after a great French soldier and army commander of the previous Great War. It was destined later to become closely associated with Western Deep Levels.

CHAPTER FIVE

". . . Deep levels are getting fashionable . . ."

(Lionel Phillips in a letter to Messrs Wernher, Beit and Co., 3 October 1891)

In 1945 peace came in two stages: firstly, in May with the conclusion of the war in Europe, and then in September with the end of hostilities in the Far East. South African forces had been involved in both: more deeply in the former than in the latter.

By then an unnamed female had earned dubious fame by driving an infinite number of teams of six white horses round endless mountains throughout the Allied world, had worn inestimable pairs of silk pyjamas, and had further performed endless unmentionable deeds while careering around and around those mountains *en route* to her unspecified destination.

Vocal reminders of a white Christmas were cursed in Europe's icy winters and, with nostalgia, loved in the tropical jungles of the East and the sandy, brown deserts of Africa.

In those years, too, World War I's sweetest girl from Tipperary had been replaced — for the Springboks — by their own musical heroine, a beauty who hailed from the maize lands around the Mooi River in the western Transvaal.

Men — and women too — exchanging their uniforms for "civvies" again, returned to their peace-time occupations. But not all who had gone to fight came home again. Among the nation's warriors were numbers whose bravery was acknowledged by medals and decorations. South Africa had reason to be proud of its sons and daughters who had served the cause of freedom.

The nation's industry, knee-haltered during the war by shortages of materials and of labour and also by the diversion of work from peace-time norms to essential military functions, was champing at the bit to return to normal production. But the time in between had not been wasted, and, in the longer term, the knowledge gained was to prove useful. From that war-time experience, South African industry acquired valuable expertise. This its technicians and planners developed and built upon. It was to stand the country in good stead years later when, its contribution to the Allied cause seemingly forgotten, South Africa came to be threatened with economic isolation.

As great as the incentives were for all, perhaps the gold mines had reasons more valid for wishing to get on with the task of building the future. During the agitated war years there had quietly been exciting developments in South African mining, progress which held out splendid prospects for the years ahead. Dr Krahmann's introduction of the magnetometer to geological surveying in South Africa a decade and a half before had revealed fresh sources of underground wealth waiting to be exploited: gold-fields waiting to be tested and tapped. Mining houses, realising the importance of the need for sound financial resources in South Africa's post-war role, had been planning the development of new properties, especially in the Free State and along the West Wits Line, to be undertaken as soon as times and circumstances became propitious.

Ironically, Dr Krahmann — to whom the South African gold mining industry owed so much — had been interned as an enemy alien at the outbreak of war. Soon the South African government repatriated him to Germany, where, in Berlin, he was captured by the Russians and spent several years in a concentration camp. Released in 1949, he returned, a sick man, to South Africa, where he died in 1960.

While hostilities interfered severely with Anglo American's plans for the exploitation and development of the new mining areas becoming available, planning and preparation for the post-war future had prudently continued. This foresight enabled the Corporation to minimise delays when the war ended.

Priority was given to developing the Free State gold-fields so that it was only in the fifties that the exploitation of the area south-west of Randfontein, along the West Wits line, really got under way.

In 1950, at a ceremony at the Welkom Mine in the Free State, Sir Ernest said many people regarded the discovery of the Free State goldfields as extremely fortunate and opportune for South Africa's economy. "The Rand is replacing itself", he declared, and pointed to the development of new mines on the East and on the West Rand. Further, he hinted at "other expansion programmes" under consideration which showed that the Rand itself was "dealing with the replacement of its so-called 'dying mines'."

Since the formation in 1944 of its Western Ultra Deep Levels, Anglo American had been giving considerable thought to the possibilities — and problems — of mining at depths never before penetrated. Research had revealed the presence of gold in the "lost" reef which beyond Randfontein dipped, it was now discovered, to depths of approximately 10 000 to 12 000 feet (3 100 to 3 720 metres). Operations at such depths, hitherto untouched, would involve techniques hitherto unused. There was much to think about. . . . What were the limits to which mining could reach? The question was by no means a new one. It had been debated from the beginnings of underground mining, with increasing intensity as greater and greater depths were reached. On the Witwatersrand the American mining engineer, John Hays Hammond, in the 1890s had realised that the dip of the reef would necessitate deeper and yet deeper operations. He was confident that shafts could be sunk to depths of 4 000 feet (1 240 metres).

The engineers rose to the challenge. Time extended the limits, and sixty years later in the 1950s there were four mines on the Witwatersrand working below 8 000 feet (2 480 metres), as well as another two whose workings reached to more than 7 500 feet (approx 2 330 metres) underground.

In 1925 the Government Mining Engineer had warned that the lives of fourteen producing mines on the Far East Rand were running out. In 1943 the Minister of Mines set up a Committee again to examine the whole question of deep-level gold mining. Its Chairman was Sir Robert Kotze, a distinguished former Government Mining Engineer who had told the authorities eighteen years previously that the lives of those mines "on the Far East Rand, estimated upon a conservative basis, range from three to twenty-three years and average eleven years."

With gold mining the mainstay of the nations's economy, the South African government — and, indeed, the whole nation — continued to have a very real interest in the prospects of the industry.

With earlier theoretical limits now far exceeded, however, practice was once again challenging theory and threatening to batter even the freshly conceived frontiers of mining at depth.

After lengthy study, the Committee in January 1945 reported its findings. It said: ". . . Because of noteworthy advances in ventilation and in general improvements in mining technique, some of the depths which have been reached exceed the limit placed on mining operations not many years ago. It now seems, however, that without further great improvement we are very near the final limit that can be reached in existing circumstances and with present-day methods."

In delving into the earth, what were the limits of human endurance in terms of the debilitating heat encountered at great depths? How far down could mining extend and still remain economically viable? What mechanical means were required to ensure the safety of workers so far underground? Could "existing circumstances" and "present-day methods" be extended and improved? These questions — and more — were central to the whole issue of the new possibilities which faced mining houses in Johannesburg.

The Committee reported that the general opinion at which mining could be carried on profitably set limits at about 9 000 feet (about 2 800 metres) and that, "in any case, even with improvements in technique and practice, it would not exceed 10 000 feet. The Committee posed the question: "what of the ore that lies at greater depths? . . ."

What, indeed, of the ores that lie at greater depths!

What, for example, of the ore in the Carbon Leader on the West Rand, known to be lying at depths in excess of 10 000 feet (3 050 metres)? For some years Anglo American had been giving serious consideration to the problems connected with its extraction. While the Corporation was not alone in such studies, it had been the first to examine the possibilities of mining at these depths and, similarly, was the first mining house to take positive action to implement plans for such deep-level mining.

Following the decision to create Western Ultra Deep Levels came the further decision to purchase options on some 16 000 hectares to the south of the Blyvooruitzicht and West Driefontein mines. In a memorandum drawn up in September 1948, Dr Austen Bancroft, Anglo American's Consulting Geologist, described this as "the most promising area" for the starting of "a new venture based on ultra deep mining". He wrote of "the wisdom of the original plan to form an ultra deep mining venture with unequalled possibilities."

He proceeded to describe the geology of the area:

> "Southward from the Mining Lease Areas of Blyvooruitzicht and West Driefontein, the surface rises at first gently and then abruptly for a total of 300 feet to 400 feet vertically, thus forming a prominent ridge trending approximately east to west.
>
> The whole of the areas are occupied by shales and quartzites of the Timeball Hill Series which at constant horizons have been invaded by two sheets of dolerite. It is the upper sheet of dolerite which caps the ridge trending E. to W.
>
> With dips varying from 3 degrees to 8 degrees southward the

Timeball Hill Series rests unconformably on the Dolomites, the junction being marked by a variable thickness of chert breccia. To date, ten drill holes have penetrated Timeball Hill Series varying from 411 feet to 1 228 feet in thickness.

This cover of Timeball Hill Series on the Dolomites is a fortunate feature from a mining point of view. Protected from weathering influences, the dolomites are free from caverns and water channels and the accumulation of large and small chert boulders to varying depths that have too frequently made the sinking of shafts difficult and costly in areas where dolomites are exposed at the surface.

The Dolomites have an average thickness of about 4 000 feet with a dip of 4 degrees to 5 degrees southward. The base of the dolomites is marked by the conformable Black Reef Series of shales and quartzites which have an average thickness of 40 feet. Nine boreholes have failed to disclose a payable Black Reef at the base of this Series. At a depth of 4 348 feet Hole U.D.4 encountered the Black Reef assaying 11.83 dwts over 5 inches. At 4 717 feet, Hole U.D.6 penetrated Black Reef assaying 2.64 dwts over 5.5 inches. At 5 128 feet in Hole U.D.9 the Black Reef was represented by a line of pebbles within a length of 4 inches of core which assayed 2.13 dwts. In the other six boreholes no Black Reef was present.

The base of the Black Reef Series marks a very great unconformity. Prior to the advance of the sea within which the Black Reef Series and the Dolomites were deposited, there was a very prolonged period of dry land conditions during which erosion removed many thousands' feet thickness of strata.

Without expanding on this phase of the geology of the areas, it will suffice to point out that during this period the Ventersdorp lavas were completely removed from the Blyvoorzicht Mining Lease area and from more than one half of the West Driefontein Mining Lease. Where these lavas are absent, no Ventersdorp Contact Reef can possibly exist . . ."

The memorandum went on to describe how test-drilling had established the position of the sub-outcrop of the Ventersdorp lavas which, thickening towards the south-east, have an average dip in that direction of 17 degrees. Drilling in the ultra deep areas had penetrated the base of the lavas at depths varying from 4 988 feet to 5 479 feet. The same drilling found the thickness of the lava to vary from 104 feet to 2 380 feet.

There were indications, Dr Bancroft continued, "that the Ventersdorp Contact Reef may be an asset of considerable importance in the ultra deep areas . . ." and he went on to emphasise the point:

> ". . . The so called Ventersdorp Contact Reef was developed by rivers and streams flowing on the land surface upon which the Ventersdorp lavas were poured forth. Such an origin implies the deposition of gold-bearing gravels in stream channels and restricted alluvial flats. One must not gain the impression that the Contact Reef is a blanket reef that extends over the whole area at the base of the lavas. Only about 20% of the lava contact in the Venterspost Mining Area has yielded payable reef. Nevertheless, this reef must be regarded as an asset of very considerable importance in the ultra deep areas. . . ."

Bancroft concurred in the views expressed by the official Committee on Deep Level Mining in its report of January 1945.

Regarding the West Wits area, that Report had stated:

> ". . . The Blyvooruitzicht Mine has shown phenomenal values and if the property to the dip proves to have similar values, even if not so high, it could stand a heavy outlay in capital expenditure and a high working cost and still prove remunerative. In respect of this and similar areas we are of opinion that the best interests of the State are served by the opening of new properties, where feasible, rather than by extending the life of existing mines. . . ."

Addressing the Corporation, Bancroft had written: ". . . the area to the south of the Blyvooruitzicht and West Driefontein Mining Lease Areas is the most promising area on the Rand for the establishment of a new and highly profitable deep mining venture. . . ."

By July 1950 the geological and technical investigations of the West Wits area had been completed and the stage was set for further developments. With the Free State gold mines successfully launched, Anglo American was now able to focus its attention on the development of this area.

In December 1956 Western Ultra Deep Levels, acting in association with other interested companies, applied for a mining lease over an area of about 4 767 morgen — the equivalent of 6 863 claims — immediately south of Blyvooruitzicht and West Driefontein mines, as Dr Bancroft had recommended.

After lengthy delays, the wheels had at last begun to turn.

Something of the eagerness to get on with the job is reflected in a letter which Sir Ernest Oppenheimer wrote to his son, Harry, in Parliament in Cape Town in February 1957. In regard to Western Ultra Deep Levels, he said: "I spent the weekend studying the reports because I think we should come to finality this week. I hope on Friday when you are back. I drew up a memorandum . . . I enclose a copy and would like you to think it over. I personally feel that we should conclude the business with the Leases Board and so secure our position as a big gold producer indefinitely . . ."

On 6 May Western Ultra Deeps was advised by the Secretary for Mines that the Minister had agreed to the grant of a mining lease. This was to be ceded to Western Deep Levels, a new company established for the purpose.

Soon afterwards, on 6 August, Western Deep Levels Limited was legally and officially incorporated. Its objectives, specified in its Memorandum of Association, were to:

(i) adopt the flotation agreement providing for its establishment, entered into between the companies associated with the venture;
(ii) take cession from Western Ultra Deeps, Limited, of the mining lease, which the Honourable the Minister of Mines has agreed to grant that company, to mine precious metals over the area of approximately 4 767 morgen;
(iii) take cession of the mineral rights within the mining lease area; and
(iv) carry on gold mining operations in the mining lease area.

In terms of the lease the company was required to carry on shaft sinking and development in the mining lease area until an amount of £5-million had been expended, provided that operations might, with the consent of the Minister of Mines, be terminated at any prior time.

The Company was registered with an authorised capital of £851 000 divided into 850 000 "A" shares of £1 each together with a thousand "B" shares of similar value. The purpose of creating "B" shares in the initial authorised capital was to establish a distinction between "A" and "B" shares from the start and to confer upon the "A" shares a preferential right to receive dividends amounting, in total, to ten shillings per share before any dividends were paid in respect of any other class of share in the capital of the Company.

The flotation agreement provided for the payment by Western Deep Levels of £600 000 in cash to certain of the companies associated with the venture by way of vendors' consideration for the acquisition of the mining lease and for the simultaneous subscription for cash by the companies of 600 000 "A" shares. The agreement also provided for the

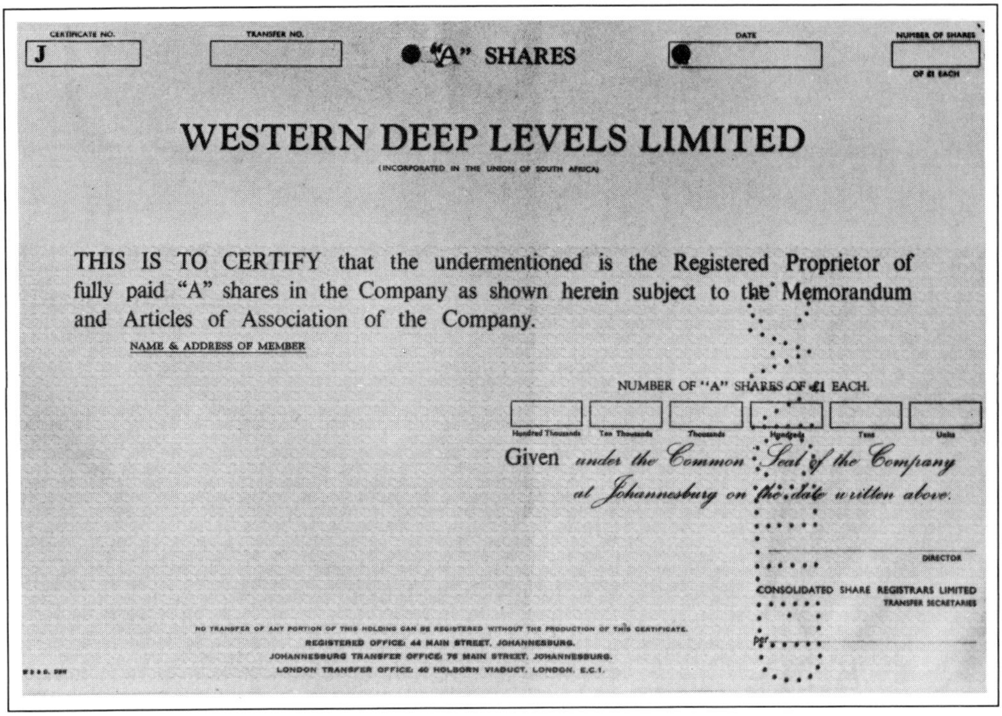

CERTIFICATE NO. J | TRANSFER NO. | "A" SHARES | DATE | NUMBER OF SHARES OF £1 EACH

WESTERN DEEP LEVELS LIMITED

(INCORPORATED IN THE UNION OF SOUTH AFRICA)

THIS IS TO CERTIFY that the undermentioned is the Registered Proprietor of fully paid "A" shares in the Company as shown herein subject to the Memorandum and Articles of Association of the Company.

NAME & ADDRESS OF MEMBER

NUMBER OF "A" SHARES OF £1 EACH.

Hundred Thousands	Ten Thousands	Thousands	Hundreds	Tens	Units

Given under the Common Seal of the Company at Johannesburg on the date written above.

DIRECTOR

CONSOLIDATED SHARE REGISTRARS LIMITED
TRANSFER SECRETARIES

per

NO TRANSFER OF ANY PORTION OF THIS HOLDING CAN BE REGISTERED WITHOUT THE PRODUCTION OF THIS CERTIFICATE.
REGISTERED OFFICE: 44 MAIN STREET, JOHANNESBURG.
JOHANNESBURG TRANSFER OFFICE: 76 MAIN STREET, JOHANNESBURG.
LONDON TRANSFER OFFICE: 40 HOLBORN VIADUCT, LONDON, E.C.1.

Specimen "A" share certificate. Western Deep Levels Limited was registered in 1957 with an authorised capital of £851 000 divided into 850 000 "A" shares of £1 each, together with 1 000 "B" shares of similar value.

subscription for cash by Western Ultra Deep Levels Limited of the remaining 250 000 "A" shares.

To provie capital funds necessary to launch the new mine, Anglo American Corporation undertook, under the flotation agreement, to form a syndicate which would lend the company £5-million initially required. The flotation agreement then contemplated the issue of 4 750 000 "A" shares of £1 each as additional funds were required. After the funds were exhausted, "B" shares of £1 each would be issued to finance the balance of the capital needed to take the mine to production.

The loan syndicate was formed by Anglo American in association with Central Mining Finance, Blyvooruitzicht Gold Mining Company, New Consolidated Gold Fields, West Witwatersrand Areas, West Driefontein Gold Mining Company and Western Ultra Deep Levels.

The terms of the loan were generous. It was free of interest for the first seven years (i.e. until 26 September 1964) after which Western Deeps would have to pay interest at the rate of five per cent per annum. The loan had to be redeemed in full within twenty years.

Besides the details about geological conditions garnered from exploratory drilling, conditions at the neighbouring mines, Blyvooruitzicht and West Driefontein, provided considerable further information about conditions underground.

The Report of the Consulting Engineer, dated 6 December 1958, set out details of the original prospecting programme undertaken in the lease area of 4 767 morgen over which the new company of Western Deep Levels had the exclusive mining rights.

Of the ten boreholes originally sunk, nine had been completed. Of those, four intersected both the Ventersdorp Contact Reef and the Carbon Leader Reef, while two intersected the Ventersdorp Contact Reef only and three the Carbon Leader Reef. The tenth was abandoned.

Results varied widely from place to place. There was nothing unusual about that.

The deepest, Borehole UD.9, for example, which went down 6 900 feet (approx 2 140 metres) yielded only a "trace" of gold from the VCR, while UD.3, drilled only to 5 000 feet (1 550 metres approx.) found ore values of 22.65 dwt. The reef there was narrow, however, so that overall its total potential content was not as rich as the reef encountered by UD.10, further south, where the dwt value, though far less, was increased by the greater width of the ore-bearing reef.

Results from the boreholes drilled to the Carbon Leader within the Western Deeps lease area showed richer prospects, at greater depths.

The figures — seen in conjunction with operating results on the adjacent mining properties to the north — were promising enough to "justify the conclusion that precious metals exist in payable quantities within the lease area." The Consulting Engineer pointed out, however, that the Ventersdorp Contact Reef, while carrying good values of gold in certain areas, "is liable to be somewhat inconsistent", with the result that the economics of mining the reef could not be estimated with any degreee of confidence.

The development and exploitation of the mine was planned to cover three stages. The first would see development of the twin vertical shafts of both Numbers 2 and 3 Shaft systems for opening up as far as the Ventersdorp Contact Reef. A reduction plant with a nominal capacity of 80 000 tons of ore per month was envisaged. The total capital cost of that stage was estimated at £19 600 000.

The second stage embraced the extension of the two shaft systems to the Carbon Leader Reef and the expansion of the reduction plant to a nominal capacity of 120 000 tons a month. This would be further increased as the mine developed. The cost of this stage was estimated

Western Deep Levels mine hospital.

at £9-million, which included the cost of some second-phase developments actually executed in the course of the first stage.

The third — and final — stage of planning envisaged expansion to attain the full productive capacity of the mine; and an extended reduction plant aimed at handling 200 000 tons of ore per month.

As soon as the grant of a mining lease was assured and with the assurance that funds would be available, preparatory work began on the site towards the end of 1957. Early the following year shaft sinking began with Number 3 ventilation shaft. Before the middle of 1958 work had begun on the sinking of the Number 3 main Shaft and, at the Number 2 Shaft complex, on the sinking of both the ventilation and main shafts.

The taxation of South Africa's gold mines has been described as "ingeniously devised". Perhaps that is so. Certainly interpretation of the laws requires ingenuity, for the system is highly complex and generally felt, even by tax experts, to be something of a treacherous minefield for those who have to traverse it. Nevertheless, it is a very fair system.

In terms of the mining lease granted, the considerations payable to the Government of the Union of South Africa were: (a) a percentage of the annual profits expressed by the formula $Y = 15 - 90/X$, where Y is the percentage of profits payable to the Government and X is the ratio of profits to recovery expressed as a percentage; (b) an amount equal to one and a quarter per cent of the amount determined in accordance with (a) above; (c) a rent of six shillings per month per morgen or fraction thereof included in the mining lease area.

In addition, taxation imposed by statute on gold-mining companies was payable on the profits of the company after deduction of the lease consideration.

In view of the fact that the new Western Deeps gold mine had been especially planned to mine at deep levels — deeper than any previously — the government granted tax concessions to ease the burden of the additional expenditure in terms of development costs made necessary by these extraordinary operations.

This necessitated the formulation of a definition of what constituted "deep level". Since the first discovery of gold on the Witwatersrand the early concept of "deep level" had developed and changed. No longer did it mean merely any mining operations underground. To meet the evolution of those changing concepts, in 1945 the Committee on Deep Level Mining had suggested that "for the purposes of this report the words "deep level" will denote a depth below 8 500 feet."

That committee had further suggested that "the taxable profits would be reduced by the amount of capital spent on the deep level portion" of development and, too, that "it would be open to the company to put forward any formula or combination of formulae that it likes" for the assessment of taxation in relation to the particular circumstances of that mine.

It was in the spirit of these recommendations that the new Western Deep Levels now acted.

In seeking tax relief for the obvious — and substantial — additional costs involved in deep mining practices, the first Managing Director of the company, John Shilling, consulted the Government Mining Engineer about ways and means of handling the additional financial burdens. The GME expressed understanding of the problem and showed sympathy for the proposal for some special consideration. He said, however, that he could not single out Western Deeps for special treatment and pointed out that by its very nature the request really fell within the ambit of the Income Tax Act. This therefore put the matter beyond the scope of his particular office.

John Shilling was not a man to be deterred.

At once he took the train to Cape Town, where Parliament was in session, to discuss the matter with the Minister of Mines, then Dr A.J.R. van Rhijn, who listened sympathetically to Shilling's request for tax relief commensurate with the additional financial burdens which the newly established mine would have to bear. Dr van Rhijn suggested that his visitor might care to set out a form of words which might be considered for inclusion as an amendment to the Income Tax Act.

The invitation was swiftly accepted. Shutting himself in a room at the De Beers offices in Cape Town, Shilling drafted terms he felt would meet the situation. To start with, this necessitated a definition which at the same time would meet the unique requirements not only of Western Deep Levels but also of other deep-level mines, obviating any possible discrimination which the Government Mining Engineer feared.

The challenge of finding an appropriate formula was well within Shilling's competence, for he had been trained in law and had come into the mining industry as the legal adviser in the Chamber of Mines.

Seizing on the concept described by the Committee on Deep Level Mining in 1945, he suggested that tax relief be granted to mines which operated below 8 500 feet. Because the circumstances were new, he added the qualification that such a concession should apply to new mines — those with leases issued after 30 June 1956.

In general the government accepted these recommendations, and the Income Tax Act of that year included a provision for an amendment to the principal Act (No. 31 of 1941) to accommodate the situation facing the new "deep level gold mines".

The provision read as follows:

> ". . . 'deep level gold mine' means any new mine in respect of which —
> (a) the Governor-General or the Minister of Mines, as the case may be, on the recommendation of the Mining Leases Board has, after the thirtieth day of June, 1956, signified in writing his decision to grant a lease of the right to mine for gold; and
> (b) the Government Mining Engineer is satisfied that, at the time of the decision to grant the lease is signified, the principal object is to mine gold bearing ores believed to exist within the lease area at depths below a vertical depth of seven thousand five hundred feet from the surface; and
> (c) the Government Mining Engineer is of the opinion, at the time of the decision to grant the lease is signified, that a period of at least seven years will elapse from the date upon which shaft sinking excavations commence to the date of deep level production."

At last "deep levels" had been legally defined. It was indeed appropriate that Western Deeps should have pioneered the crystallising of the concept — and the baptising of the term!

For the newly established Western Deep Levels this concession meant the addition to redeemable capital expenditure of an amount equivalent to five per cent per annum compound interest on the unredeemed balance of capital expenditure. This was calculated up to the end of the year immediately preceding the first year of assessment in which the company's aggregate net profits from the commencement of production exceeded whatever capital expenditure had been incurred since the company's inception.

Early steps in the formation of Western Deep Levels were proving successful. With pleasing rapidity things were falling into place: doubtless a consequence of good planning. Test drilling had provided encouraging basic information. The necessary ground had been acquired. The statutory mining lease had been obtained. Initial venture capital was forthcoming in the form of loans from the parent Anglo American Corporation. Negotiations were coming to a successful conclusion and the new company was swiftly gaining an identity and independence of its own.

Barclays Bank D.C.O. became the company's bankers. The Anglo American Corporation took over the functions of administrative and technical advisers as well as those of secretaries to the new company; Alex Aiken and Carter were nominated as auditors, while Messrs Webber, Wentzel, Hofmeyr, Turnbull and Company were retained as legal advisers and solicitors. Consolidated Share Registrars, Limited, were listed as the company's transfer secretaries with effect from 1 January 1959.

With the passing of almost 30 years, Barclays Bank has undergone some changes: links between the parent bank in the United Kingdom and its South African affiliate were broken in 1987, when funding by Anglo American enabled the South African institution to become independent. This then became the First National Bank of South Africa. As such it continues to operate as bankers to Western Deep Levels. Similarly Anglo American continues as technical advisers and secretaries. Alex Aiken and Carter, the company's name also slightly modified, has continued without interruption as the auditors, while the link forged at the outset with the Consolidated Share Registrars has not been broken.

Little did those first workers at the mine site realise how soon the mielie fields they were clearing in the shallow valley of the Gatsrand area would be replaced by one of the richest mines in the world — and technologically the most advanced. Nor could they have pictured how, within a few years, the bare veld there would be covered with an assortment of buildings and machinery which would transform the quiet rural area into a dynamic industrial complex of very considerable importance to the country's prosperity.

The 24th July 1957 was a sunny, winter's day; one of those days when the brightness of the sunshine on the Transvaal highveld gives a deceptive impression of the temperature. At Western Deeps a keen breeze added a further chill to the cool air as it stirred up wisps of dust from the earth where in recent weeks the soil and the tawny grass of the veld had been ground to fine powder by heavy trucks. On veld that had been bare a few weeks previously, clusters of work sheds were now scattered, while piles of materials here and there, where once had been mielie fields, indicated that a vast amount of construction lay ahead. Events were creating a momentum of their own.

A small crowd gathered at the site of the Number 2 Shaft of the new mine to see Sir Ernest Oppenheimer inaugurate an enterprise that was destined to launch a new era of mining, marked by operations to be taken to depths hitherto untouched anywhere in the world and at depths to which man had never before ventured.

For Sir Ernest it was the end of the beginning in the development of a venture close to his heart, one which had owed its evolution to his personal vision, courage and conviction and to his determination which had not wavered, even when some of those around him were possessed by doubts.

Sir Ernest told the press that negotiations for this project had been in progress for fifteen years. But if for some this was the culmination of years of patient planning and organising, for others it was a beginning — the start of operations and of exciting new challenges.

On another occasion in Johannesburg in that same week, Sir Ernest said that the new Western Deep Levels mine "... is the most important enterprise ever started in South Africa ...", pointing out that, when it reached production, it would produce as much gold in a year as the entire industry in South Africa was then producing in a month.

In those years of conception and planning, with patience and foresight, Sir Ernest had carefully brought together the threads necessary for the implementation of the enterprise. Not merely a new mine — another gold mine — Western Deeps brought a whole new dimension to mining. Each of the collaborating team in his own way had played a vital and important part in the development of the whole. Bancroft's geological expertise and advice had led to the selection of the site on the West Wits Line, while the skills of the Corporation's senior consulting engineer, Frank Unger, had indicated the means of exploiting its wealth and developing ways of extracting it. R.B. Hagart, close confidant of Sir Ernest and a Deputy Chairman of the Anglo American Corporation, had patiently arranged the purchase of the ground and, with similar patience, negotiated the complicated financial agreements with other mining houses with which the Anglo American group had had to deal.

Dr H.C.M. Whiteside, a distinguished Consulting Geologist whose long service with Anglo straddled World War II, reminiscing a few years ago about early days in the Corporation, said that Western Deeps had largely been designed by Bancroft and Hagart. ". . . if it had not been for those two," he added, "we would possibly not have a Western Deep Levels . . ."

In the final analysis, however, it was not least the tax concession granted by the government in respect of new deep level gold mining which made the gigantic new venture financially feasible and attractive. This concession was not only solid encouragement to a flagging mining industry, but was also a very real and practical financial inducement for the mining industry as a whole and of considerable benefit to the country. Sir Ernest was generous in his expressions of gratitude for the assistance which the government had given.

Johannesburg's *Star* described the venture as a partnership between the State and private enterprise. By implementing ideas put forward by the Committee headed by Sir Robert Kotze, the State "had conceded the principle that those financing large enterprises should be allowed to recover their capital wholly or largely before the State takes a share of the profits."

The mine was expected to have a life of between fifty and sixty years. At the time of the symbolic inauguration of its operations, the press speculated that in this time it would produce gold to the value of £800-million, but of course that estimate did not — could not — take into account the frequent fluctuations in the value of the metal. Journalists also estimated that the mine would cost some £20-million to develop to the stage of production, six or seven years ahead, but once again that figure took no account of economic inflation in the future.

The *Rand Daily Mail* pointed out that Western Deep Levels "will eventually be the deepest, most costly and largest gold mine in South Africa." The scale of investment — most of it raised in the country — made it a symbol of confidence in the future of South Africa. Another newspaper wrote that: ". . . as long as South Africa can find the men with the faith, the initiative and the daring to grapple with such problems as have already been solved by her mining and financial leaders, there are no limits to what the future could hold."

At the inaugural ceremony in July, Sir Ernest said, "although I am almost 78 now, I hope I will be able to see the first gold that is brought up from this mine."

But, sadly, that was not to be. He died in November, a few months after launching this great new enterprise. The establishment of Western Deep Levels was the last major mining project with which Sir Ernest was concerned, while the simple ceremony at the planned Number 2 Shaft was one of the very last public functions he attended. Writing about his father several years later, Mr Harry Oppenheimer said: "... The opening of a new mine at depths of 10 000 feet and more below the surface was something new in mining history, and Western Deeps was planned, from the start, on an exceptional scale. It was a project which involved risks from which many younger people in our organisation recoiled. But my father, in his early seventies, became chairman of the new company, occupied himself with the details of the financial planning involved and looked to the future of the mine with complete confidence and tremendous enthusiasm."

There were indeed several in those planning days who recoiled from the risks which they felt would be involved in pushing mining to these new and untried limits. And it was Sir Ernest who acted intuitively — as he had done before — to take the decisions and to bear the risks and responsibilities which others were inclined to avoid. But, as his son has pointed out, those risks were carefully calculated. Though apparently spontaneous, none was haphazard. HFO goes on to tell how Sir Ernest considered major decisions "were best taken by some intuitive process, though he realised that these intuitions came only after you had been deeply immersed in the stream of affairs and had worked hard to understand them ... What my father called intuition was really an ability to make up his mind quickly about the probabilities of a case and to act unhesitatingly on his judgment. ..."

Three days after the formal incorporation of the company on 6 August 1957, the first meeting of its Board of Directors took place at Anglo's headquarters at 44 Main Street in Johannesburg.

Locked away in a cavernous strong-room in the offices of Anglo American's Gold Division are the Company's minute books; thick tomes, heavy, solid and austere. Their firm bond pages, cold to the touch, are bound between sombre red leather, the covers, embossed in gold, bearing the dates of each. Dignified and conservative in that fashion which tradition demands of records of such significance and importance, they are made yet more venerable by a clinging musty aroma characteristic of archival records kept constantly under lock and key.

The minutes of that first Board meeting were typed on an unusual typewriter, an antiquated-looking instrument which remains a historical curiosity in the Corporation's archives. It is an archaic machine which types directly on to paper positioned horizontally and which can be used to type on pages of books of varying thicknesses placed beneath it.

Present at that historic first Board meeting were six directors: Sir Ernest Oppenheimer, together with W.M. Frames, E.S. Hallet, P.S. Hammond, J.W. Shilling and T. Reekie. Apologies for absence were accepted from R.B. Hagart and W.D. Wilson.

The Minutes record the Board's first Resolution thus: "Resolved that Sir Ernest Oppenheimer be appointed Chairman of the Board of Directors of the Company." The decision was unanimous.

It was the only meeting of the Western Deeps Board over which Sir Ernest presided. He was unable to be present when the Directors convened again in September and in October. In his absence, John Shilling, who had been appointed Managing Director of the company, then took the Chair.

South Africa mourned the loss of a great man when Sir Ernest died. His contribution to the country's development had been immense. The minutes of the Western Deep Levels board meeting on 3 December record his death and "The meeting stood in sympathy."

At the mine the initial work was progressing apace. The contract for the design and construction of four shaft collars and headgears had been awarded to the Roberts Construction Company, which published a periodic newsletter to report progress on the work. The newsletter of 28 November expressed the sadness of those who had known and worked with the late Sir Ernest Oppenheimer, and mourned the death of a great South African. Recalling the simple ceremony he had performed to set in motion the pre-grouting drills at the official start of work, the newsletter spoke of "the great hopes he had for the future of this mine". It went on to say: "History will recall the plain facts of the achievements of this great man, who for close on fifty years has played a major part in the mining and industrial development of Southern and Central Africa. From the human point of view, however, probably his greatest monument is the happiness of the millions of people in many parts of the world who are directly or indirectly employed by the ventures which have prospered as a result of his foresight and financial genius. In addition, generations yet unborn will benefit from his remarkable vision and from the many institutions of learning and charity which were established as a direct result of his keen sense of justice and compassion for the common people . . ."

Mr Harry Oppenheimer succeeded his father on the Board of Directors and in January of the following year became Chairman. He held the position until December 1962. By then Western Deeps was in every way an independent concern, established and productive. Though stepping down from the Chair, Mr Oppenheimer continued to serve as a member of the Board until 1 September 1979. In those years he saw, inevitably, numbers of directors and alternate directors come and go, some to other positions, others away to retirement, some to situations elsewhere. Among those on the original bench of Alternate Directors was Mr Gavin Relly, destined later to become the distinguished Chairman of the Anglo American Corporation.

At the formal opening of the mine in March 1962, John Shilling spoke of the part which Mr Oppenheimer had played in the initial development of Western Deeps. He said: "As you will appreciate it is not possible, in a Group such as ours, for the Chairman to take charge personally of many individual projects or to give his attention to the wide variety of detailed decisions that have to be taken from day to day. In this case, however, Mr Harry Oppenheimer, on his father's death, agreed to lead the project himself. The task of finding some some R60 000 000 to meet the bills constantly being turned in by the consulting engineers has fallen to him and his associates. His imperturbable spirit in finding the finance and in meeting changing circumstances and his qualities of leadership and decisive analysis of problems have been our sheet anchor and our inspiration. . . ."

With the complexities of the initial financing and formation of the Company now finalised and with the tempo of the early operations at the mine site accelerating with the passing days, the focus of attention passes from the Company's board room and administrative offices to the activities at the mine itself. There the air was electric with the excitement of the challenge.

CHAPTER SIX

Meanwhile, back at the Mine . . .

Anticipating the completion of the legal and financial formalities for the establishment of the new mine, a provisional organisation consisting of technical staff had been set up as early as April 1957 to co-ordinate the planning and the initial stages of the mine's development.

The functions of the committees and working groups covered all aspects of the initial operations, matters such as the surface layout, location of the various buildings, pre-cementation in connection with the shaft sinking, the design of hoists and permanent headgears, as well as studying the designs of hostels, the location of housing and recreational facilities. Infinite challenges faced the planners.

With the entity of the company established, the provisional status of those preparatory committees fell away and their functions were fused into an established operational organisation. There was no hiccup and no time was lost. The work went on smoothly as the provisional planning meshed into the confirmed arrangements. Planning moved into overdrive. In the first intra-office memorandum of the new main organising committee, the Chairman announced that the Manager of the new mine and certain other appointments had been made and said that, for the most part, the mine now fell to be dealt with in the ordinary way.

The establishment of mines of the calibre and importance of the new Western Deep Levels is by no means an everyday affair. Positions in organisations of such prestige are highly regarded and much sought after, and understandably there were many contenders for each prestigous post. Every candidate had his backers. Each contender had outstanding qualifications of the sort that the responsibilites and initiatives and challenges of the new venture demanded. From many candi-

Number 3 Shaft complex at the mine set close to the northern slopes of the Westwitsrand.

dates, Frank Ludwig was chosen as Manager. At the time of his appointment, he was Manager at Anglo's Springs Mine.

Frank Ludwig had walked a long road in the mining world; from working at the stope face through many stages of experience to management level. At the time of his appointment at Western Deeps he was described as "fresh-faced, calm, authoritative," and certainly he took his responsibilites with the seriousness they demanded.

The appointment was confirmed, with effect from 6 August 1957, the day that Western Deeps was legally incorporated and from the Registrar of Companies received its formal "Certificate to Commence Business".

At the same time the Board appointed Bert Hulley as Mine Secretary "at a salary of £130 per month plus cost of living allowance." Over 40 000 cheques, representing millions of pounds as well as rands, passed through his hands in his first five years at Western Deeps.

Frank Ludwig's function, wrote the *Rand Daily Mail,* was "... to translate a vision and a blueprint into reality..."

And this he certainly did.

He steered the destiny of Western Deep Levels from the beginning of the great enterprise, not only until the mine reached production in 1962, but also for a number of years after that when, having been promoted to General Manager, he left Western Deeps at the end of January 1966, for reasons of health, to take over the management of East Daggafontein. He had seen the Western Deeps enterprise grow from open veld to "one of the largest and most modern" mines in the world, and in doing so he built a bridge to a new era in mining.

In his valedictory address to the staff at the mine, Frank Ludwig insisted that one man could not, and did not, make a mine and emphasised that the success of Western Deeps was due to a team effort, and that he had been blessed with "a team second to none". All had pulled their weight and, while it was invidious to single out individuals, he did wish to pay special tribute to three of his staff whose assistance and advice had been particularly invaluable: the first two were Walter Wise and Dick Lambert. By virtue of their positions as Resident Engineer and Assistant Manager, respectively, they had been key figures in the early planning and establishment of the mine. To these names he added that of Don MacIver, who took over from Lambert when the latter was transferred. To Barry Nieuwoudt, Regional Organiser of the Mine Workers' Union, he also paid special tribute, saying that there was "only one thing on a mine worse than a fire and that was a breakdown in industrial relations."

In 1957, before the formal commencement of operations, not only had working groups been established, but the preliminary tasks of site

Surveyors at work on the site of Western Deep Levels mine: 1957.

clearance had been undertaken; while, in collaboration with the geologists, the precise location of the first two shafts had been decided upon by the Consulting Engineer, "Nobby" Clark. Roberts Construction had been awarded a contract for constructing four shaft collars and the concrete headgears, and it was envisaged that collar construction, shaft sinking and the building of the headgears, as well as several operations, would be executed "in rapid succession, if not simultaneously". The closest co-operation among the various participants, supervised by the Anglo American Consulting Mechanical Engineer, Desmond Bentley, was essential. That such co-operation was readily and fully given from the start soon became apparent from the whirlwind pace at which the work progressed. A long haul lay ahead; but despite the daunting extent of the distance into the future, the arduous marathon began — and continued — at a sprinter's pace.

Initial planning envisaged two shaft complexes, each comprising a twin-shaft system. There were several sound, practical reasons for adopting this design and, additionally, the ventilation shaft provided each complex with the second outlet which mining law prescribes.

In each complex the main hoisting shaft and the ventilation shaft were situated 61 metres apart. It was planned that the main shaft should be sunk to approximately 2 000 metres below surface to exploit the Ventersdorp Contact Reef; and, while that was being worked to provide an initial cash flow, sub-vertical shafts would then reach down to about 3 150 metres below surface for the mining of the Carbon Leader. The ventilation shafts, however, were planned to reach down in single, continuous vertical shafts to the lower operating level, a novel idea made possible by the Blair stage winders which saved enormous amounts of time — and costs.

Throughout those early stages, planning was assisted by scale models of various sections of the mine. These three-dimensional models visually reinforced the infinite number carefully drawn plans. Western Deeps was fortunate in having the services of two brothers, Alf and Hermann Seelos, both expert model makers, the excellence of whose work may be admired from the models preserved in the Mining School at the University of the Witwatersrand. Frank Ludwig once said that the use of those models had saved Western Deeps "every cent of a million rand" in time and planning.

Progress up to 9 September is described in detail in the first report submitted to the Chairman by the new Consulting Engineer, Harry MacConachie. This covers not only progress in the areas of both Number 2 and Number 3 Shafts, but also on various essential buildings:

Early construction work at shaft head.

the offices and store-rooms, staff quarters and workshops. The pace of achievement continued to be impressive.

Statistics tell the story. During August a total of approximately 1 040 metres of pre-cementation work was accomplished in ten diamond-drill holes. A week later that figure had been extended to approximately 1 570 metres, and a total of 13 374 pockets (of a little more than 45,5 kilos each) of cement, slag and fly-ash had been injected into water-bearing fissures. Collar excavation had been completed and so had concreting to a depth of about 15 metres at Number 2 main shaft and also at Number 3. The foundations for hoists and headgears were well under way and the first store-rooms and site office were nearing completion.

The impressive pace of the developments at "the world's most fantastic embryo mine" was commented on towards the end of August by a columnist of the *Rand Daily Mail*. "Today," he wrote, "I saw the rolling hills south of Carletonville blitzed into the beginnings of a throbbing metal and concrete complex. A desert road of red dust, fine as face powder, led to a scattering of prefab buildings, store sheds built between a Wednesday and a Friday, the skeletons of engineering shops, contractors' compounds, and a sprawling range of works incomprehensible without a guide and explanations. ... Steadily diamond drills plug downward ... ton after ton of cement is poured into the earth around the course the shaft sinkers will follow to a depth below at the height at which the average air liner cruises over South Africa. ..."

Clearly the magnitude of the project itself posed a challenge — and that challenge was a spur to progress. Every week, almost every day, the face of the mine site altered as new buildings mushroomed from the earth, as fresh access roads were laid and the veld became increasingly scarred as heavy trucks churned through the dust or the mud, which ever the fickle spring weather dictated. Workers, sensing the challenge, responded splendidly. Towards the end of November, Roberts Construction reported that its men "... have been working all and every day and most nights as well, in a continuous effort to keep abreast of the programme. ... Going one better than the proverbial candle, they are burning this one in the middle as well as at both ends. ..."

The periodic schedules of progress testified to that claim.

The rapid rate of achievement in those initial months led Desmond Bentley in December to issue a message to all workers:

"... As we hope to be sinking the first shaft before the year end, I want to take this opportunity, before the Civil Contractors' Christmas break, of conveying to you my personal appreciation and that of the Corporation for the tremendous effort and long hours of hard work that you have put into the project, in order to bring Western Deeps to its present state of development. Taking all the circumstances into consideration, this has been achieved in record time. At the beginning of April, when the project was released, not a single drawing had been prepared and we had no equipment other than the two MV 4700 H.P. hoists. You will recollect that I called our first meeting with Roberts Construction Company on the 16th April, at which I outlined my preliminary ideas regarding the design of the main and the ventilation shaft headgears, and the construction programme to be followed. Further meetings with Roberts Construction Company's engineers followed in quick succession, and work started on the site on 15th June. In the short space of time of some eight months we have acquired and are erecting no less than ten permanent hoists. ... In the meantime, Roberts Construction Company have virtually completed the main winder houses, the compressor houses, and the ventilation shaft headframes: the collar and the preliminary shaft sinking and lining at both collars; the No. 3 Main Shaft headframe to the sinking height of 135 feet (approx 42 metres); the shaft collar, shaft sinking and lining, etc., together with a great deal of work on complementary services. ... Permanent housing, offices, stores and workshops have been built. ... This magnificent progress could certainly not have been achieved had it not been for the fine team of engineers in the Roberts Construction Company's organisation and in the excellent co-operation which has been established between the various companies, the Management and Staff of the mine, and the Corporation. ..."

The hope of "sinking the first shaft before the year end" was not realised, however; but that work commenced early in 1958.

To reach the Ventersdorp Contact Reef, shafts had to penetrate three main geological strata — all, of course, with varying degrees of thickness and hardness. On the Western Deeps property these, in descending order, are what the geologists now call the Pretoria Group, which range between about 185 and 470 metres in thickness. Below that is the Chuniespoort Group, varying between about 1 100 and 1 425 metres in depth. Not only does the hardness of this dolomite present drillers and shaft sinkers with immense problems, but the formation conceals cavities into which water has seeped through the ages. These vast underground reservoirs present an extreme hazard to any shaft sinkers who trespass into those areas. Finally, before reaching the Ventersdorp Lavas, shafts have to penetrate the Black Reef Formation — in this region estimated to be between about 10 and 12 metres thick — consisting of quartzites and carbonaceous shales and frequently containing large nodules of pyrite.

Geologists describe the Ventersdorp Supergroup itself as consisting of amygdaloidal lavas of uneven thickness. On the Western Deeps sector, the thickness increases towards the south and east parts of the property, and the reef dips at an angle of about 20 to 25 degrees.

Sandwiched between the Ventersdorp Lavas and the rich Carbon Leader Reef are two further distinct and well-defined geological strata. Immediately below the Ventersdorp Lava lies the Turffontein Subgroup, with formations consisting of quartzites, conglomerates and shales, with little trace of gold. The Kimberley Reef, however, is not present across the entire mining area of the Western Deeps property and is absent from the region to the east, where Number 3 Shaft is situated.

Below the Turffontein strata, nature has created the Johannesburg Sub-Group, interleaving — as it were — the two auriferous strata which lie at these great depths.

The Carbon Leader is situated between 75 and 80 metres below the Middelvlei Reef and some 600 to 1 200 metres deeper than the VCR. Deriving its name from the association of a basal seam of carbon with the thin overlying conglomerate, this reef lies along an axis that points roughly 75 degrees north-east and with a fairly constant dip, from north to south, of 21 degrees.

This information, fundamental to the evaluation of the mine, had been patiently pieced together from years of exploratory drilling, the cores minutely examined by geologists over many years of deep probing. The data were no less important to those implementing the decision, for planning had to take into account not only where the gold was located and at what depths, but also to consider the requirements of machinery, of staffing as well as a hundred and one related matters.

Those considerations would be extensive enough in an ordinary mine, and there was no lack or shortage of experience of what was required for that. But the extraordinary depths which faced planners at this new, ultra-deep mine brought a host of additional problems, some of a sort that had never been tackled before.

The primary aim of a commercial enterprise is, of course, to ensure that earnings exceed expenditure, and in this sense gold mining is no exception. In mining, however, the calculations are complicated by the number of variable factors and risks are thus correspondingly higher. There is, for example, no guaranteed consistency in the values of the ore being mined: furthermore, there is no guaranteed price for the product, since the selling price all over the world can — and does — yo-yo, up or down, often under the influence of some wholly unexpected political occurrence in a remote corner of the globe.

Mining skills must essentially be supplemented by management expertise as well as by the acumen of cost accountants and planners.

R.N. (Dick) Lambert, Assistant Manager at Western Deeps in those early days, has explained the extent to which depth brought greater challenges and problems. He has spoken of the "aggravated problems of finance, rock temperatures, rock pressures, water and the ever-lengthening lines of communication which materially affect the transport of men and material as well as the transmission of power. Coupled with this is the steady whittling down of that vital portion of the shift spent on actual production . . ." These problems are common, of course, to every mining project but in this mine are exacerbated and intensified by the increased depths to be traversed, so that, in even greater measure, their solution, is "... a prerequisite to efficient mining..." Repeatedly he emphasised the importance of pre-production planning.

Dick Lambert's experience in mining and underground construction added authority to his words. He had worked in copper and asbestos as well as in gold mines, broadening his experience in Rhodesia and Ghana as well as in South Africa.

Now retired, he still recalls with obvious pleasure the challenges and grinding hard work of those early years of Western Deeps and describes them as "the most dynamic period of my working life." He has said, too, "... what a wonderful opportunity it was for those of us who were sent there to plan and bring the mine into production... It was an opportunity seldom given to mining men..." But of his wartime experiences in the 1940s he remains modestly silent. It was then that his long-held beliefs in the value of planning and in the importance of time and timing were reinforced. They were inherent in his contribution to a vital and urgent part of the Allied war effort in Italy.

The South African Corps of Engineers, with specialists in several technical fields, played a significant role in the Second World War; and because their work was often essential to the forward momentum of the battle, while clearing mine fields, reconstructing bridges and making roads, the sappers came under enemy fire almost as often as the front-line infantry.

The specialist knowledge of the South African miners among the Engineers on more than one occasion proved invaluable, and it was in the clearing of a sabotaged railway tunnel in Italy that Lieutenant Lambert's mining experience contributed to the speed of the Allied pursuit of the retreating German forces between November 1944 and May 1945.

The Company of South African Engineers in which he was serving was faced with the problem of restoring two railway tunnels through the Apennines which had been destroyed and blocked by the Germans in their northward retreat late in 1944. The use of these tunnels was vital to the Allies' offensive planned for the spring of the following year. That volume of South Africa's official War History devoted to the Corps of Engineers, *Salute the Sappers,* describes the re-opening of the Serravalle Tunnel particularly as "a high priority task".

The tactical significance of these tunnels had been fully appreciated by the Germans, and they had done all possible not only to render them useless and impassable, but also to make any quick restoration difficult. The portals had been blown to smithereens. In the middle a massive demolition blockage reinforced the outer obstructions. The sabotage had been carefully planned and skilfully executed, leaving any restorers to face an awesome challenge. But the enemy had reckoned without the ingenuity of experienced miners, men with practical knowledge of underground tunnelling. *Die boere het 'n plan gemaak!* Quickly a makeshift hoist was fabricated and erected over a vertical ventilation shaft. This was used to lower 200 men a day into the heart of the smashed tunnel and return them to fresh air at the conclusion of a shift. This it did thoughout the reconstruction work without a single mishap.

It was an ingenious contraption. The shaft, Lambert explains, was equipped with a 65 hp diesel-driven docking hoist with two drums in tandem and capstan fitted on both ends of the drum shaft. The clutch was between the hoist and the motor. Pull was in one direction only, so that all the lowering was done on the brakes, which needed frequent attention, and during heavy hoisting periods jets of compressed air were necessary to keep the brakes and drums cool enough to operate at all. The headgear was built of light steel "scrounged" from stores.

The clearing and restoration of these tunnels by Springbok Sappers has been officially described as "an amazing piece of work carried out under very difficult and hazardous conditions". It has been said, too, that "the urgency of the military situation called for the utmost skill and speed . . ." and in the case of the Serravalle Tunnel South African Engineers completed the job within one hour of the time originally estimated.

The war over, it fell to Dick Lambert's lot as Assistant Manager in the early days of Western Deeps to be in control of the planning section.

A paper he prepared in October 1959 on the way in which the initial stages of the mine's development had been handled demonstrated the extent of the detail with which every phase of those early operations had been planned and implemented. The subsequent success of Western Deep Levels — and the margins of profit — testify to the soundness of that preparation and of the teamwork, both in the offices and at the site.

While the shaft sinkers continued busily grinding their way downwards with the hoisting and ventilation shafts, new buildings were being added to those already on the surface and planning went on.

Leal Wright, Western Deeps' first Underground Manager, headed a committee "to carry out an extensive investigation of current stoping methods with the object of developing suitable methods for stoping under deep mining conditions." In the course of its investigations, the Committee visited 25 mines and studied their stoping methods so that designs for Western Deeps would be based on the latest techniques of shaft layout and stoping in the exploitation of both the Ventersdorp Contact Reef and the Carbon Leader.

After careful consideration, it was decided that longwall stoping had several advantages over scattered mining. This was the system then adopted and developed. Dimensions of the individual stopes were adapted to match the sizes of the payable areas they served.

Rock temperatures are a key factor in deep-level operations. They present problems in all mining, but the intensification of heat with increasing depths becomes a headache for planners and engineers who have to deal with it. These temperatures present a physical barrier to labour efficiency; overcoming them becomes a factor in the economic viability of the mining operation.

Being the first mine planned to operate on such a large scale at depth, Western Deeps has had to pioneer techniques to overcome the excessive heat problems so far underground.

J. de V. Lambrechts, Group Ventilation Engineer of Anglo American, was engaged in 1959 in studying conditions likely to be encountered at Western Deeps. He said that "with the accumulated knowledge and experience gained in establishing deep mines, ventilation planning for Western Deep Levels mine presented a unique opportunity for applying that knowledge and experience from the grass roots down, as it were."

As far back as 1949, in anticipation of the problems of high temperatures likely to be encountered deep underground, studies were instituted to gain information about virgin rock temperatures at various depths. This was done by measuring the temperatures in the deep boreholes being drilled in the area south of the Blyvooruitzicht and West Driefontein mines, where Western Deeps in due course came to be located. These investigations also showed temperatures on mines in particular areas to be roughly similar. Rock temperatures on the Free State mines are higher — depth for depth — than those on the Witwatersrand and in the western Transvaal.

Armed with this information, engineers were able in advance to plan the ventilation requirements for the embryo mine at Western Deeps. In addition to the large quantities of surface air which have to be pumped underground, it was found that cooled air was essential in working areas. This led to the installation not only of huge fans, but also of giant refrigeration plants to chill air, to cool water and to produce large quantities of ice, needed underground to cool not only workers, but also much of the machinery. Practical experience at the mine has continued to further knowledge of conditions and requirements underground.

The capital cost of this equipment is extremely high, and the costs of its installation, running and maintenance add considerably to the cost of gold production.

By the middle of February 1958 — a little more than six months after Sir Ernest's symbolic inauguration of the mine, *The Star* reported that concrete headgears "the height of 12-storey buildings" now stood where ten months earlier there had been only bare veld. The headgear for the Number 2 Shaft had been raised to 135 feet in a mere nine days, a South African record for the "sliding" method of construction.

At Western Deeps square concrete headgears took the place of the metal A-frame style of headgear which had hitherto generally been used on gold mines in South Africa and which had become traditional landmarks on the highveld. These towers were designed by "Chuck" Rigby, a specialist in reinforced concrete design, who later became a Professor of Engineering at the University of the Witwatersrand.

The shape and style were new, though Western Deeps was not the first mine to use these rectangular towers which, although sheathed in concrete, still contained enormous amounts of steel in their construction.

Dick Davidson, appointed Chief Engineer at the mine in 1960, relates the story of how Felix Parry, then Chief Surveyor at Western Deeps, was puzzled to find, when the winding equipment was being installed, that these towers appeared to cant. Careful instrument readings from time to time showed that they were not always truly perpendicular — as they had to be for the winding machinery. He felt the builders to be at fault, as instrument readings, precisely and meticulously recorded, continued to indicate some lateral movement.

The fact that the buildings leaned consistently one way in the mornings and slightly in the opposite direction in the afternoons provided clues to the puzzle. It came to be realised that the hot, strong sunshine caused the construction to expand more on the sunny side than on the side in cooler shadow, so that the building canted slightly, one way or the other, from the perpendicular. The necessary modifications were quickly made!

By the end of 1958 the loan funds available to the company had been exhausted. Capital expenditure on shaft sinking, buildings and equipment from the start of operations to the end of December had amounted to £5 482 907. Capital expenditure for 1959 was estimated at £333 600. The excess of expenditure over funds available had been financed by a short-term loan from Anglo American. The time had arrived to raise additional capital through the medium of the "A" shares.

In terms of a prospectus issued on 30 January 1959, the company made an offer for subscription in cash at par of 4 750 000 shares in the capital of the company. The offer was underwritten and fully subscribed.

The company's first annual general meeting took place at 44 Main Street on 19 May 1959. It was presided over by Mr H.F. Oppenheimer whose review, as well as the reports of the Directors and the Consulting Engineer, covered the period from the incorporation of the Company to the end of 1958. The Chairman was able to tell shareholders of the "good progress . . . being made on the property."

This was fully supported by the Consulting Engineer who gave details of achievements at both shafts; though at the Number 2 complex there had been nagging problems with water-bearing fissures, the sealing of which had required additional cementation treatment.

Harry MacConachie, Consulting Engineer during the initial operations, had been appointed to the Board and had been succeeded as Consulting Engineer by Barrington (Barry) Mudd, a graduate of Witwatersrand University and a civil and mining engineer with a distinguished war record in the South African Corps of Engineers.

The South African war history recounts how in North Africa Lieutenant J.B. Mudd, in charge of a section detailed urgently to clear a road through a minefield for the passage of urgent supplies, "himself undertook the more hazardous tasks" in clearing an unusually densely mined and booby-trapped section of the obstacle. His fine example speeded up the clearance work considerably and he was decorated with the Military Cross.

Western Deeps has been well served by numbers of ex-servicemen — too many to mention — to whom battle had added maturity and a deep sense of responsibility and had taught them the richness of comradeship in service.

In Italy, by then promoted to major, Barry Mudd received the MBE ". . . for his ingenuity in the design of the special articulating pier supports and reinforcing bridge members manufactured from standard Bailey equipment". This bridge, set up in amazing quick time across the broad River Po, was named "Springbok Bridge" and was described as "one of the greatest achievements" of the South African Engineer

Corps. The bridge — and the amazing speed of its erection — contributed substantially to the hastening of the Allied victory in Italy. Its conception and construction were the brain-child of Barry Mudd who now, as Anglo's Consulting Engineer, became responsible for bringing Western Deeps to production.

Gordon Henderson, Consulting Engineer involved in the early planning of Western Deeps, had also served with the Springbok Sappers in North Africa where he had been rewarded for his courage.

Sent out one night with a section to clear a path through a minefield, Lieutenant G.H. Henderson's "coolness and quick decisions brought his section safely through two tank attacks . . ." The War History goes on to recount that "His personal disregard for danger and sound judgement under intense fire saved his men from destruction more than once. His conduct earned him the Military Cross for gallant and distinguished service . . ."

It began to look as if Western Deep Levels had been properly launched and, with a momentum of its own, was now charting its own course. Of necessity the voyage ahead would be a long one and — almost certainly — all would not be wholly smooth sailing.

John Shilling, Managing Director, had earlier outlined a proposed scenario for the immediate future. ". . . with an undertaking of this size," he had said, "it is most necessary to view the progression of the mine in phases which could be efficiently controlled by a continuous review of estimates and construction techniques . . . the first important phase is that which it is expected that the Ventersdorp Contact Reef will be intersected in either one of the ventilation shafts. In future reports to the Board this will be referred to as the first phase. The subsequent period which will involve bringing the mine to a milling rate of 100 000 tons per month will be referred to as the second phase . . ." He went on: ". . . A tentative date for the completion of the first phase has been fixed at 30th June, 1959, and it is at present estimated that . . . an amount of £7 390 000 will have been spent on shaft sinking and equipment."

That scenario was being duly implemented.

The initial work on the collars of the four shafts completed and the foundations of the surface headgear under way, the thrust of the major work passed from Roberts Construction to the shaft sinkers.

In those tasks then facing the embryo mine, Western Deeps was fortunate in having the services of some of South Africa's best and most experienced shaft sinkers. Among them, to tackle the new Number 3 Shaft, was Chris van der Westhuizen, Master Sinker, who came to Western Deeps from Vaal Reefs where he had "broken all records in sight", and whom Barry Mudd described as "one of the country's greatest Master Sinkers" and "an outstanding and magnificent leader of men". With him, and following him later, were men of similar competence and unusual ability, such as Jim McAlpine, Keith Richards and E.R.Brune. They had also set world records.

The adjacent Number 2 Shaft encountered greater problems than its twin to the east and it speaks volumes for the ability of Pieter Swart, Master Sinker in charge of those operations, that delays were minimised. Fifteen times as much fissure water by volume faced workers on Number 2 Shaft than on Number 3.

As fast as the shafts were sunk, so with similar speed did the surface buildings stretch upwards. In the drive downwards, the friendly rivalry between the teams at the respective shaft complexes egged the sinkers to strenuous and yet more strenuous efforts. And progress was made

Shaft sinkers going underground in a kibble, led by Chris van der Westhuizen, Master Sinker and world record breaker shaft sinker.

manifest in other ways, too: by the growing numbers of workers and the increase in all branches of the staff and, no less, by the houses and hostels which were being built for them. New roads were being opened across the mine's property to facilitate the expanding range and size of traffic bringing men and material to the scattered work sites. *The Star* in Johannesburg described the scene as "controlled chaos" with "a creeping jungle of power cables, thicker than a man's wrist . . ."

At each complex the shaft-sinking shift changed every six hours, as men, chasing the gold that lay thousands of feet away, continued to sink what was destined to be "the deepest hole in Africa".

Towards the end of that second year, to expedite longer-term development and preliminary investigations of the Ventersdorp Contact Reef, an approach was made to the neighbouring West Driefontein mine, proposing that certain developments on Western Deeps be carried out from West Driefontein. It was planned to drive into the Western Deeps lease area from Level 16 of the neighbour's Number 5 Shaft, nearest to the common border.

By cross-cutting for some 2 500 feet (775 metres) to the boundary and then continuing for a further 300 feet (93 metres) at that level, such a probe would enter the VCR on Western Deeps property. This development could give Western Deeps contact with the VCR well in advance of either of its own shafts, enabling the mine, through this co-operation, to undertake a number of developments through that area, not least in regard to test assays to estimate values and also to prepare for the mining of the VCR when its own Number 3 Shaft reached there.

At the Company's second AGM, held in May 1960, the Chairman, reporting on the progress of operations during 1959, the year under review, touched upon this project, telling shareholders that the enterprise had unfortunately not yielded the assay results that had been expected. "The development has been disappointing," he said, pointing out, however, that experience gained elsewhere in mining the Ventersdorp Contact Reef had indicated that enrichment occurred in large zones separated by areas of low payability.

The fierce pride which miners have in their work and the intense loyalty to their mine leads, inevitably, to a great sense of friendly rivalry. In addition, managements nurse a certain confidentiality about their intentions and aspects of their operations. But at the same time, this individuality and loyalty in no way impair co-operation between mines or among miners whenever the need arises. This was demonstrated by the ready collaboration between the members of separate mining groups in the project that enabled Western Deeps to explore its share of the VCR through the developed shafts of its older neighbour to the north.

This general co-operation was again demonstrated when, shortly after taking over as Consulting Engineer at Western Deeps, Barry Mudd sought the co-operation of engineer colleagues from a number of other mining houses. Inviting twelve of them to visit the Western Deeps site, he asked them, after inspecting its early developments, not only to comment on the planning, but also to offer criticism and point out any faults or weaknesses. At the same time he canvassed suggestions for improvements. In this way, the advice of the country's experts was enlisted to ensure that the costly enterprise was on the right track.

He confesses that he was especially concerned about the manner in which the VCR would be exploited at Western Deeps. The rock in that formation was hard and brittle, with little elasticity. This accentuated the risk of devastating rockbursts and, Mudd admits, "that worried the hell out of me."

"That conference in June 1960 showed," he said, "how, in our planning, we tried to solicit, as widely as we could, ideas from throughout the entire mining industry. This meeting was a good illustration of the broadness of the co-operation we sought to ensure that we were going about this job in the best possible way."

Shaft sinking continued at a cracking pace throughout 1959, workers at the two shaft complexes battling through the 1 550 metres of tough dolomite. On 11 February 1960 the Number 3 Main Shaft reached its final planned vertical depth of 1 970 metres, and nine days later its companion ventilation shaft reached 1 960 metres where further sinking was halted for the time being, until the onward drive to the Carbon Leader — almost 930 metres further down — began. At that time, Number 2 Shaft, hampered by additional problems of its own, was about 280 metres from its target.

With the attainment of these initial objectives, both main shafts were stripped of all sinking equipment and work began on the installation of the permanent hoisting machinery in readiness for the exploitation of the VCR.

The speed of achievement — not only underground but on the surface as well — forced planners early in the following year to revise their estimates and schedules. The press speculated that milling would commence significantly sooner than had been expected with, of course, consequently significant benefits of lower capital expenditure.

There had been investigations and consultations with engineers in

June 1960: Visit to Western Deep Levels of Consulting Engineers from other mining houses at the invitation of Barry Mudd, Consulting Engineer to Anglo American supervising operations at Western Deep Levels.

(Left to right) Front row: *John Irvine (Rand Mines); Bill Clucas (Union Corporation); Francis Hill (Rand Mines); Frank Ludwig (Manager; Western Deep Levels); Barry Mudd (Anglo American Corporation); Reg Cousens (Gold Fields of SA); J. B. Bocock (Gold Fields); Dick Cooke (JCI).*

Back row: *Dick Lambert (Assistant Manager, WDL); Neil Holford (Anglo Transvaal); Bernard Roberts (General Mining); Clifford Linsell (General Mining); Jeff Goode (Union Corporation); Buster Fowler (Rand Mines).*

several countries about hoisting equipment suitable for working at such depths. Barry Mudd and others had been overseas in search of technical advice which led to the decision to instal Koepe winders at both main shafts. It was broadly agreed that this equipment would be able satisfactorily to handle hoisting from the great depths being pioneered by Western Deep Levels. One engineer — a German — expressed doubts as to whether the equipment would be able to cope with operations from such depths. This lone opinion was to prove of significance a few years later.

During shaft sinking, the necessary pumps and underground ventilation equipment had been installed at the various levels. Plans envisaged an ultimate pumping capacity of some 30-million gallons of water per shaft per day. And while water was being removed from underground, a large reservoir for the storage of water required for men and machines had been constructed on the surface.

March 1960 saw the first exposure of the VCR on the Western Deeps property when the Number 2 Main Shaft intersected the reef at a depth of 1 735 metres below surface and yielded more encouraging values than those in the cross-cut operation from West Driefontein in 1959.

Pay gold at last !

Soon afterwards the ventilation shaft struck the reef at a similar depth. At Number 3 Shaft, however, the reef lay more than 125 metres further down. It was not until December, after further sinking, that the ventilation shaft there intersected the reef at a depth of 1 957 metres.

This was a significant milestone along the way. But by no means was it the end of the road.

By the end of 1959 a fresh charge of capital funding became necessary. A total of 2 800 000 "B" shares of £1 each were offered for subscription, part to "A" shareholders and part-members of the loan syndicate at a price of 40 shillings per share, thus raising a further £5 600 000. Each of these "B" shares carried the right to subscribe for further "B" shares of £1 at a price of 45 shillings per share. In April 1961 the price had fallen to R3 per share (the currency having been decimalised with R2 equalling £1), because of the deterioration in market conditions. The shares under option were offered for subscription with a further issue of 2 800 000 "B" shares at the same price.

With fresh capital, the exploitation of the VCR began: and work went ahead above and below ground to bring the mine to production on the VCR. At the same time, a start was made with the further sinking of the vertical ventilation shafts towards the Carbon Leader horizon, while preparations were also set in train for excavations for the sub-vertical main shafts down to that richer target.

Inevitably, such concentration of industry had changed the face of the environment. Large buildings and tall towers dominated the landscape, while bald, expanding waste dumps, mimicking nature's ancient koppies, began changing the topography of the veld. The bed of the valley became criss-crossed by roads and tracks, while houses sprouted from the slope of the southern ridge. Here was a new world, a place that was now never without lights, nor ever without the rumble of activity.

But the impact upon the environment of this sudden change was unfortunately not limited to the landscape. Towards the end of 1960, in the adjacent valley to the north, a number of houses in Carletonville began to show signs of cracking. At once the authorities, properly concerned, invited Professor Jennings of the Witwatersrand University to investigate the causes of the problem. He reported that subsidences were occurring as a result of the lowering of the water table in the area. He warned that further earth movement could be expected. Asked to undertake further investigations he was consulted on the siting of various mine buildings.

By that time, Western Deep Levels had built — or with subsidies that had enabled employees themselves to build — more than 130 houses in Carletonville. The company had also built a number of flats. But, after studying Professor Jennings' report, the Board decided to find alternative sites for mine housing.

Walter Wise, Resident Engineer at Western Deeps, was living in one of the mine's first houses in Carletonville when the cracking began. The company had built a number of attractive houses in Aster Drive for senior personnel. Four of them, adjacent to one another, were occupied by Frank Ludwig, Bert Hulley, Piet Swart and himself. One after the other, the buildings began to crack badly, to the point where daylight could be seen through the walls. Patching up proved useless and one by one the houses were abandoned and demolished.

The Wise's home was still intact, however, when, in 1960, he was transferred to Head Office in Johannesburg. A year later, while on a visit to the mine, Walter Wise and his wife went to see their previous home. "Our three neighbours' houses were gone," he relates, "no bricks, no foundations, no fences — not even a sign of there having been a garden. Our house was standing vacant. The driveway to the garage, instead of having a rise of about two feet, was about two feet lower. It was an eerie experience walking about the garden . . ." Nostalgia drove them to look at other parts of the town. "We came to a barricade," he says. "Looking ahead we could see the rooftops of six or seven houses at road level."

Sinkholes — as the term implies — occur when the ground suddenly subsides into cavities below. They are generally caused by the withdrawal of water or other liquid (such as oil) from beneath the surface.

Sinkholes are not a modern phenomenon: they have occurred throughout the world for many millions of years, though occurrences have become more frequent since man began tampering with the earth.

The dolomitic nature of the Wonderfontein Valley makes the West Rand area — in which Western Deeps is situated — particularly vulnerable to these sudden subsidences. The word Gatsrand, "ridge-with-holes", itself indicates natural cavities there and is descriptive of the nature of the earth in that region.

The government initiated investigations into the problems of these sinkholes. Reassurance came some years later when those studies had been completed. At the company's AGM in May 1965, the Chairman explained the situation to shareholders: ". . . In recent years," he said, "the company has been closely connected with the investigations into the occurrence of ground movements in dolomitic areas on the Far West Rand. The safety of the underground workings and surface installations on the company's own mine is not in doubt as the company's property is located on thick and impermeable geological strata which protect the underlying dolomites from water percolation, and the likelihood of subsidence or collapse is remote. The company is, however, affected both financially and administratively by the occurrence of sinkholes in the Carletonville area.

"First, in the interests of the greatest possible safety of its employees and their families, a number of the company's mine houses in Carletonville have been vacated and demolished as a precautionary measure. As much as possible of the salvaged material has been used in the erection of replacement houses, the additional cost of those so far built being R430 000. Secondly, arising out of the decision of the Government not to allow surplus water pumped from the mines in the Oberholzer and Venterspost dolomitic water compartments to be returned to these compartments, the company has joined with other mines in the area in a voluntary association, the Far West Rand Dolomitic Water Association, to restore water supplies and to settle claims arising from damage or loss which is directly attributable, as the proximate physical cause, to the implementation of the policy of dewatering . . ."

Suddenly — without warning — the full extent of the danger of sinkholes was made manifest.

On the morning of 12 December 1962, Western Deeps' northern neighbour, the West Driefontein Mine, lost an entire screening, sorting and crushing plant which vanished in a trice into a vast sinkhole about 50 metres across and 30-odd metres deep. The earth had suddenly crumbled and opened up to swallow the entire plant, burying it in an enormous cavity. At the same time, the main pipe carrying the underground water up West Driefontein's Number 2 Shaft was severed and immediately water began pouring into the gigantic sinkhole at the rate of a million gallons an hour. Rescue operations were rendered impossible and twenty-nine lives were lost in this extraordinary and horrifying accident.

The situation called for immediate action if further subsidence of the side-walls were to be prevented and if further damage and loss of life were to be forestalled. The risks were enormous. None knew whether there would be further subsidence. At once workers began dumping waste rock into the gaping hole, as fast as the risky circumstances allowed, never sure how near to the edge of the hole they could safely venture with their equipment, never sure how stable the surrounding area really was. Fears remained unspoken. On and on they toiled through each day and through subsequent weeks. By 8 January, in less than four weeks, the gaping hole had been filled, some 175 000 tons of waste rock having been pushed and poured into it. The story of the rehabilitation of the site, the interim operations of the mine and its return to production is a shining example of team-work, ingenuity and planning. The cost of restoration amounted to about R9-million.

For South Africa 1961 was a particularly significant year. It was the year in which South Africa left the British Commonwealth and in which it became the pioneer in decimalising its sterling-patterned currency. A South African pound became equivalent to two rands. The switch to decimals was later followed by others — Australia and New Zealand on the same basis and later by Britain itself. Throughout the year progress continued at Western Deeps. Developments devoured money, however, and several times in that year recourse was had to further loans and to further funds raised through increased shareholdings.

The Bird Reef was intersected in April. It had not been expected that this reef would prove rewarding, and assays confirmed the correctness of the assessment. The series of reefs, carrying "four well-developed conglomerate bands and seven pebble and grit stringers", was "found to carry negligible values", as the Chairman was to report to shareholders. It was of no economic interest.

Test milling began at the mine in November 1961 and the extent of mining operations, well ahead of schedule, soon made it apparent that the reduction plant, planned for the milling of 80 000 tons of ore per month, would not be large enough to handle the quantities of ore available for hoisting. The plant had been designed to permit expansion in units of 40 000 tons. Extensions were begun at once to increase the plant's monthly capacity to 120 000 tons. Reserves of ore taken from the VCR continued to build up so that by late November substantial quantities were being held in readiness for milling. In the same breath, as it were, engineers were contemplating further extensions to mill 160 000 tons a month. Production was racing ahead of planning.

Suddenly actual production — gold and yet more gold! — began to loom very close.

Bill Nienaber, Underground Manager at Number 3 Shaft, tells that when experimental stoping commenced in September 1961, the target was the production of 80 000 tons of ore per month by the following March. "It was estimated that 48 stopers would be needed. But we achieved 110 000 tons with only 19 stopers and, in addition, a stockpile of 200 000 tons had been built up."

In December 1961, fourteen months after the commencement of the development opening up the VCR, Shafts Numbers 2 and 3 at Western Deeps were linked when the return airway of the twin haulages on 66 Level holed-through in record time. It had entailed driving a distance of some 3 000 metres between the two shafts, as well as developing more than 7 100 metres. The two drives met exactly — their centres less than 25 centimetres out of line: a triumph for Felix Parry, Chief Surveyor, and his staff.

At Western Deeps, a quarter of a century later, they still tell the story — doubtless now embroidered by time — of how Felix Parry, nervous on that day and concerned about the accuracy of the calculations, early in the morning phoned the Underground Manager of one of the shafts and suggested that he should again check his reckonings.

"When will the holing through take place?" Parry is said to have asked. Somewhat testily the Underground Manager replied, "We here are going to hole-through today. But I don't know about the b*******s on the other side !"

Sharing in this achievement were the mine captains of each shaft: Leyton McCann and R.E. Page, as well as the Underground Managers, Arthur Comyn, Bill Nienaber and David Dawes. Manager Frank Lud-

6 December 1961: Linking of Shafts 2 and 3 on 66 Level. The first man through the gap, Frank Ludwig, Manager, greets E.D. McCann, Mine Captain.

wig was the first through the hole when, with press photographers busy, he crawled on all fours through the jagged aperture, followed by his Assistant Manager, Dick Lambert.

The tunnelling had been done non-stop from both shafts in three shifts of eight hours a day, on and on. Water fissures were encountered along the way, though none serious, and, to everyone's relief, the teams were able to report a total absence of methane gas. Assays from diamond drill holes put down from the hanging wall haulage between the two shafts gave indications of profitable returns throughout that portion of the reef.

With the mine now within an ace of production, 1961 ended on a high note for Western Deeps. The hope grew that soon the Company would be able to start contributing financially to its own further development and, probably not far hence, also to redeeming the enormous amount of venture capital which had been incurred in development.

The audited balance sheet showed that at the end of December 1961 the authorised capital stood at R46-million. The value of shares issued amounted to R28-million. To ensure the availability of capital funds to bring the mine to production on the VCR, a further six-million "B" shares were issued, after an increase in the authorised share capital in May 1962.

The anticipated capital expenditure up to the commencement of production had in fact been exceeded by some R2,3-million, while an unexpectedly early demand for housing — due to faster development than had been expected — had compelled certain expenditure to be brought forward, so that in that sector there was an unexpected deficit of some R2,4-million. Anglo American came to the rescue with an interest-free loan of R10-million to bridge Western Deeps' shortfall until profitable production began early in the following year.

But the VCR was already being mined and shafts were well on their way down to the rich Carbon Leader: great optimism prevailed.

CHAPTER SEVEN

Laurels are not for resting on !

In the early stages of developments at Western Deeps brawn seemed repeatedly to outwit brain. Workers frequently made a mockery of the planners' timetables, as time after time targets were reached ahead of schedule. The production of the first gold is an example. When, in March 1962, the mine was brought to production on the Ventersdorp Contact Reef, operations were four months ahead of schedule, and there is no doubt that this achievement acted as a powerful stimulus to further success.

In his Chairman's Review at the Annual General Meeting a few weeks later, Mr Oppenheimer surveyed the progress achieved from the middle of 1957 when work had begun. "By August, 1960," he said, "the vertical sections of the No. 2 and No. 3 Main and Ventilation Shafts had been successfully sunk through ground notorious for its water content. By October, 1960, the No. 3 Main Shaft was commissioned, followed by the No. 2 Main Shaft in April, 1961. . . . By the end of 1961 a total of 180 126 feet had been developed on the Ventersdorp Contact Reef horizon and elsewhere in the mine . . ." And he went on to give details of the progress on the surface, the construction of workshops and offices, as well as housing for the staff, both white and black.

Mr Oppenheimer reported that, up to the commencement of production on 1 March 1962, more than R55-million had been spent on capital account in bringing the mine to this stage of development.

". . . the principal object of the company when it was formed was to exploit the Carbon Leader Reef which is being mined successfully in the Blyvooruitzicht and West Driefontein mines immediately to the north. This reef enters the company's property on its northern boundary at a depth of about 7 500 feet from surface and is thought to extend down-dip to a depth of about 13 000 feet at the southern boundary. When the

Staff houses on the Westwitsrand.

concept of a deep-level mine in this area was evolved," he went on, "it was assumed, on the basis of information then available, that the area of payability of the Ventersdorp Contact Reef, which lies about 2 500 feet above the Carbon Leader, would be somewhat restricted and that, on average, high values were not likely to be encountered. It was thought, nevertheless, that the Contact Reef would make a material contribution to the economics of the mine and the decision was taken for its exploitation as a first stage in operations leading ultimately to the opening up of the lower reef. This decision has been justified by the encouraging results of development done so far from No. 3 Shaft . . ."

Five years of hard and intensive work, tackled without fanfare and with quiet confidence and determination, plus several years of earlier studies and planning at last were beginning to bear fruit. Not least for those associated with its development, the fruit looked good and tasted sweet.

The drawn-out negotiations regarding the legal cession of properties and mining rights were finally concluded, so that the company's annual report for 1961 was able to confirm that

"The company's property and mining rights consist of:

1. A mining lease over an area of 4790.2517 morgen (equivalent to 6897.9624 claims) embracing portions of the farm Blyvooruitzicht No 116, Driefontein No 113, Oog van Elandsfontein No 114, Kraalkop No 147, Elandsfontein No 115 and Buffelsdoorn No 143, all registration division I.Q. in the districts of Oberholzer and Potchefstroom.

2. The mineral rights in respect of an area over a further 313.4892 morgen, which together comprise the company's mining lease area.

3. The surface rights over 609.5448 morgen of the farm Blyvooruitzicht No 116, I.Q., district Oberholzer.

4. The company owns a number of stands in Carletonville, on the majority of which houses for European personnel have been erected."

Now producing gold, Western Deeps was taking on a profile of its own; and it was decided that the time had come for its formal opening. There had been no ceremony since Sir Ernest had set the first pre-grouting drills to work on a mid-winter's day almost five years before.

At a function on 22 March 1962, attended by some 350 prominent South African and overseas personalities, the Governor of the Reserve Bank, Dr M.H. de Kock, performed the ceremonial opening. Mr Oppenheimer, presiding, said the occasion "... marked the completion of the first stage in the development of what is probably the most ambitious gold mining project launched since the proclamation of the Witwatersrand gold-field in 1886 ..."

Introducing Dr de Kock, the Chairman went on: "... For all of us, this occasion marks the realisation of an ambitious dream, and it touches me — and indeed all of us in the Anglo American Corporation — in a special way because it is the last major enterprise undertaken by our group which was directly inspired and initiated by my father. It is fitting that it is also the greatest and most ambitious single enterprise which we have ever undertaken.

"It was typical of him that he should have devoted to Western Deeps all his enthusiasm and driving power and ability to inspire confidence in others, although he realised and often said to me that he would not see this mine as an established producer. For him, Western Deep Levels represented a contribution to the future growth and prosperity of South Africa which had no less significance to him because he knew that personally he would not share in its benefits. ..."

While Anglo American naturally felt a special pride in Western Deeps, Mr Oppenheimer pointed out that the Corporation could not claim the whole credit for what had been achieved. "It has been in many ways a co-operative effort," he said. For a start, there was the tremendous help from the Government which had made special tax concessions to encourage ultra-deep mining. Without this it would not have been possible to raise the huge capital necessary for the project.

He stressed, too, that this was not only a co-operative effort in the sense that it required the sympathy and help of the Government alone.

22 March 1962. ". . . Ek verklaar hiermee die Western Deep Levels myn behoorlik geopen. . ." Dr M. H. de Kock, Governor of the South African Reserve Bank, formally opens Western Deep Levels mine.

"We also depended on the support of our partners. From the beginning, this enterprise rested on a close partnership between the Anglo American, the Central Mining/Rand Mines and the Gold Fields groups," and Mr Oppenheimer expressed delight at seeing representatives of those groups present at the ceremony.

"Fortune favours the brave — or anyhow it ought to do so," he continued, "and on this occasion I really believe she has, for the contact reef, the upper of the two horizons which we will work, has opened up far more favourably than any of us had the right to expect. If, as I am confident will be the case, operations on the Carbon Leader — when we get to it — go equally well, we shall have established one of the very great mines of the world."

Replying, Dr de Kock referred to the Western Deep Levels mine as "the greatest and most ambitious mining project ever undertaken in South Africa" and paid tribute to Sir Ernest's courage and acumen in going ahead with a project of such magnitude undeterred by "the formidable problem of the financing of such a venture."

Describing the enormous expense incurred in the development of a project of this magnitude — approximately R56-million so far — Dr de Kock praised the faith and patience of the shareholders, and in particular the sponsoring houses which had to bear the financial brunt of the burden. All of them certainly deserved a liberal reward for the trouble and the risk which they had taken and for the enterprise they had shown.

The occasion provided him with an opportunity to pay tribute to the gold-mining industry as a whole, which he described as "the sheet anchor — or driving wheel — of the South African economy". It was not only the country's principal source of foreign exchange, but was also of great significance to the economy of the whole free world.

Dr de Kock spoke mainly in English, but when he pulled the lever to set the mills in motion, a ceremonial act officially opening the mine, he said: *"Sover ek weet is nog geen nuwe goudmyn in Afrikaans geopen nie. Ek gaan dus vandag hierdie groot en ryk goudmyn in Afrikaans open en ek wens hom alles van die beste toe. Ek verklaar hiermee die Western Deep Levels Myn behoorlik geopen."*

The gesture set a fitting seal of South Africanism on a great South African enterprise.

Amid cheers Mrs Oppenheimer, wife of the Chairman, poured the first bars of gold to come from the Western Deeps mine. As the fiery molten gold streamed from the huge crucible into the six oblong moulds, J.H. Mortimer, Acting Consulting Metallurgist, told the spectators packed into the smelthouse how the gold had been won from the

". . . Touch for luck . . ." At the opening ceremony on 22 March 1962 (left to right) Mrs de Kock; Mr Harry Oppenheimer, Chairman of Western Deep Levels; Dr M. de Kock, Governor of the South African Reserve Bank; Mrs Oppenheimer and Mr Frank Ludwig, Manager of Western Deep Levels mine, honour tradition by touching — for good luck — the first bar of gold produced at the mine.

rocky earth. As soon as the gold had cooled, the ceremony concluded with the traditional lucky touching of one of the bars. It is a mining custom. In this ceremonial act, Mrs Oppenheimer was followed by Dr and Mrs de Kock touching the luck-inducing, pristine gold bars which sparkled in the flashing lights of the jostling press photographers.

The ceremonial pouring went off without a hitch. But the trial run on the previous day had not been without unusual problems. In the usual way, the gold was melted by heat from an electric carbon stick. As this was being withdrawn from the crucible, a link snapped and the carbon element "dropped back into the pot, splashing a shower of precious molten gold all over the place," as an eyewitness described.

The commencement of production on a mine of the magnitude of Western Deeps was a newsworthy event indeed and the occasion received considerable publicity both in South Africa and overseas. It was forecast that Western Deeps was "destined to become the world's biggest, deepest and most costly gold mine" in all the world. And, one might add, perhaps the richest too. Tribute was paid — deservedly — to the men who had planned the great project and to the workers who so splendidly had made the plans come true.

Press reports told of the eight miles of shafts that had already been sunk, of the "between 30 and 40 miles" of tunnels already gouged from the earth. Newspapers reported that at its present depth of 8 460 feet — roughly 2 500 feet below sea-level — the Number 3 shaft was already the "deepest single shaft in the world", with workers already pressing on to the planned depth of 10 400 feet (3 224 metres).

The luncheon at the Goldfields West Recreation Club in Carletonville after the ceremonial opening was a splendid, colourful and happy occasion. Fine South African wines accompanied excellent fare. At each of the several tables a senior member of Western Deeps management acted as host.

Turning to one of the hosts during lunch, a guest asked when he, as a shareholder, could expect to receive his first dividend, only to have it suggested, with a pointed wave of the menu, that he was getting it then !

In a brief speech, John Shilling said "it would be wrong of me to suggest that we at Western Deep Levels have achieved something no one else has ever achieved." But, he stressed, the Western Deeps project was a large one that was destined to play a special part in the economic well-being of South Africa. This was a source of considerable pride to all those connected with the mine, and he thanked colleagues for what they had done in its creation and development.

At the Annual General Meeting in 1963, the Chairman, commenting on the previous year's progress, reiterated that not only had production at Western Deeps begun four months ahead of schedule, but also that the initial milling through-put of 110 000 tons and a working profit of R289 063 for that month constituted a record maiden tonnage and a record profit for a new gold mine anywhere.

After only ten months of actual production, in December 1962 the Company was able to declare a dividend of ten cents per share. This reflected the hard work, sound management and planning in those initial years.

From the day Sir Ernest had inaugurated operations, work had not ceased for a moment, by day or night. Along the way in those five years a number of mining "firsts" had been recorded, a number of records broken. Tribute was paid to those achievements, and laurels were heaped upon the men who had accomplished them. But . . . laurels are not for resting on! And throughout the mine workers were eager for further progress.

By now, shortly after the official opening, there were changes on the Board. Mr Oppenheimer relinquished the office of Chairman and Hermann Koch, a Rhodes Scholar who had won a Blue for athletics, soon took his place. John Shilling, who had been Managing Director since the company's inception, was transferred to run Anglo's Coal Division. Hermann Koch held the fort briefly before handing over to Douglas Hoffe. Both Mr Oppenheimer and John Shilling remained as directors. Barry Mudd was appointed to the Board while also serving as Consulting Engineer.

For Western Deeps 1962 continued to be an eventful year. In June the Number 3 sub-vertical Ventilation Shaft was the first to intersect the Carbon Leader Reef, reaching it at a depth of 9 158 feet (2 838 metres). Development was temporarily hindered by methane gas encountered in shaft sinking, confirming the need at all times to be wary of this danger.

This first contact with the Carbon Leader proved disappointing. Values averaged only 369 inch-dwt. Gloom was dispersed by hope, however, and by the firm and undiminished expectation of better results. The initial yield was considered not to be indicative of values as a whole, and optimism was reinforced by the knowledge that in the two neighbouring mines development values had soon far exceeded those obtained from the original boreholes.

It became necessary to extend the milling capacity at the reduction plant. Capacity was increased from 120 000 to 160 000 tons a month, and towards the end of the year further extensions — to 200 000 tons — were set in train. This extension was quickly completed and by March 1963 Western Deeps was milling 200 000 tons of ore a month. This was ahead of the schedule set out in the Company's prospectus. With refined gold now to sell, the Directors appointed the Chamber of Mines to act as agents and the Company also negotiated membership of the Rand Refinery.

These further developments at the mine called for the next tranche of capital. In May this had been increased from R46-million to R50-million with a fresh share issue. At the end of the year the authorised and issued capital of the Company consisted of:

5 600 000 "A" shares of R2 each R11 200 000

19 400 000 "B" shares of R2 each R38 800 000

At the end of 1962 the company had an assessed loss, for taxation purposes, of R68-million. This figure represented the extent of the capital expenditure plus the special five per cent tax allowance for deep level mining after setting off the profits for the year.

The decision to adopt a longwall system of mining both the VCR and the CLR — under conditions of high rock pressures inherent in mining at such depths — meant that there would be some dilution in the overall grade of the ore hoisted. The longwall system was adopted to eliminate irregularities in the shape of working faces and in particular to avoid the formation of remnants and to minimize the risk of rock-bursts. The nature of the system precludes selective mining and all the rock stoped in the course of the uniform progress of the longwall is sent to the mill, whatever the grade. To compensate for this, management decided to increase the tonnages of rock processed. Soon after reaching the milling target of 200 000 tons per month, Western Deeps found that 220 000 tons could be handled without difficulty. Spurred on by the prospect of such progress, the mine's engineers decided to push up the milling rate yet further — to 240 000 tons a month. More extensions to the mill were at once set in train for completion in the middle of 1966.

That it was a prudent decision to adopt the longwall system of development was confirmed by the Technical Adviser some years later. In his report for 1967, he stated that "Experience gained over the past year in mining the Carbon Leader from stopes where faces are arranged to advance in echelon to form so-called longwalls has confirmed that, provided the payability is satisfactory, this method of mining is undoubtedly the best and safest way of exploiting a deep-lying, narrow orebody where the rock strata are under conditions of high stress."

Fortunately the ore on the VCR — so far — was turning out to be of higher grade than expected. While this was encouraging, the known inconsistency of the reef made planners and speculators cautious about predictions for the future. The width of the reef varied considerably — from a few inches in some places to as much as 100 inches in others — and geological eccentricities complicated extraction in some areas, making forward calculations difficult.

The windfall from the VCR made it possible, however, for holders of "A" shares in 1963 to be paid two dividends totalling 20 cents per share.

On 22 March that year, in the western sector of the property, the Number 2 Ventilation Shaft intersected the Carbon Leader Reef at a depth of 9 159 feet (2 838 metres) and early in September the Number 2 sub-vertical main shaft intersected the reef at a depth of 9 213 feet (2 856 metres). Samples both there and around the Number 3 Main Shaft produced most encouraging values. Again, however, jubilation and optimism were tinged with caution. The intention was ultimately to draw most of the milling ore from the Carbon Leader, and every effort was made to prospect that reef as quickly as possible.

At the AGM in May 1964, in a review of progress in the previous year, shareholders were furnished with these statistics:

Shaft	Depth below surface: (feet)	Channel width: (inches)	Value: dwt/ton	Inch dwt:
No. 2 Ventilation	9 159	11.43	60.45	691
No. 2 Sub-vertical Main	9 213	8.61	180.36	1 553
No. 3 Ventilation	9 158	35.00	10.54	369
No. 3 Sub-vertical Main	9 120	27.66	39.81	1 101

The smooth progress and the momentum of successful development received a severe jolt, however, when, on the morning of 28 October, there was a tragic overwind accident at Number 2 Shaft, resulting in the death of the hoist driver on duty, G.W. Whittington. It happened to be his forty-seventh birthday. It put the shaft out of commission for seventy days, while a further fifty-four days of intensive labour were necessary to bring that section of the mine back to full production.

Structural damage was extensive and in the circumstances it was amazing that loss of life was limited to the death of the driver. The shaft's headgear and the Koepe hoist chamber were wrecked. Debris from the smashed building and equipment littered the surrounding area. "It was like a battlefield," says Dick Lambert who, in the absence of Frank Ludwig, was acting Manager at the time. And Lambert knew what a battlefield could look like!

The main hoists were out of action. At various levels underground 209 workers were trapped. Of these, thirteen workers, busy with fresh development operations, were trapped down the sub-vertical ventilation shaft on 100 Level, 2 914 metres underground. The natural rock temperature there was an enervating 107 °F, nine degrees above blood heat. Apart from the grave risk of rapid heat exhaustion, they faced the further hazards of lack of ventilation and of rising water, as 20 000 gallons per hour poured down the damaged shaft from broken pipes. Quickly it was estimated that at such a rate the water level would engulf them within twenty-four hours.

A further 108 persons were trapped on 58 Level, 1 612 metres underground; another group of thirteen was at the pump station 125 metres higher up; yet another five were stranded at the pump station on 36 Level, while seventy workers were stuck in a cage jammed at the pump station 310 metres from the surface. Others who had been underground at the time of the accident managed to climb up to safety.

Telephone communication between the surface and underground had been partly disrupted, making it impossible to ascertain whether any of those caught underground had sustained injuries — or been killed.

One gleam of light shone through the blackness of the situation. In the northerly of the two service shafts, the "Mary-Ann" — a small service lift-cage used for minor jobs and the conveyance of small groups of people — was at surface level, undamaged and immediately usable. This was luck verging on a miracle. Rescue operations would otherwise have been an almost nightmarish impossibility. But while the Mary-Ann was intact and at surface level, there was no indication as to damage lower down the shaft. It was impossible to know how far down the Mary-Ann could in fact be used.

At the first moment of the accident, alerted by the noise and the crash of the shock wave, the driver at that hoist had halted the cages then in the shaft. Undoubtedly it was this prompt action which saved the Mary-Ann and made possible the rescue work which, in the end, prevented further loss of life. The other — southerly — Mary-Ann was lodged out of reach and out of use, immobilised at 68 Level.

The nature and the full extent of the damage was unknown. It had still to be ascertained. Immediately orders were given to halt all blasting throughout the mine to avoid the possibility of gassing men trapped underground. Rock hoisting at Number 3 Shaft was brought to a stop and the Koepe winders — identical to those at Number 2 shaft — were carefully examined, not least to ensure that in the event of problems in reaching the trapped workers in Number 2 Shaft the only remaining link with the surface, the inter-shaft connecting tunnel, would not also be immobilised.

Urgency sharpened the prevailing tension. Not knowing the extent of the damage underground, planners had to consider a broad range of contingencies. Quickly and methodically, with haste but without panic, teams set about rescuing their hapless colleagues.

Sixty-two minutes after the accident rescue operations were under way. Priority was given to reaching the men trapped on Level 100 who faced the hazard of rising water. Efforts were made to restore the vital power supply underground. By a series of links through devious other lines, telephone contact was made with the men trapped at the lowest level. To everyone's relief, there were no injuries and they had adequate supplies of emergency food and water.

Gingerly in the usable Mary-Ann the rescue teams inched their way down the shaft, inspecting damage and blockages as they went. They reached the seventy men stuck in a cage at the pump station level and found to their immense relief that none was injured. Less than two and a half hours had elapsed since the accident.

Hours ticked tensely by. That evening, a little more than eleven hours after the accident, contact was made with the thirteen men trapped further down at 1 550 metres. By 10.10 p.m. all eighty-eight persons who had been trapped in the upper sections of the shaft were safely back on the surface. But 121 men were still underground. Later that night, as emergency power cables were installed, pumping began in the main shaft. Operations deeper down the sub-vertical shaft were resumed. Within minutes of power being restored the thirteen men on the threatened 100 Level were back on the surface. The rescue work continued. Throughout the night every effort was made to get the remaining 108 men from 58 Level to the surface. Sixty-one of them had been lowered by rope down the ore-pass to Level 61 and from there brought to the surface. It was a dangerous operation, however, and one better avoided. The remainder were lifted out in the Mary-Ann as soon as they could be reached.

The whole rescue operation was completed in twenty-seven hours.

None of the miners had been injured and, it was soon seen, none was any the worse for the ordeal, though for some the experience scarred the memory, and David Grobbelaar, now a Project Officer in the Rockburst Science Division at Western Deeps, says that his hair turned grey very soon after his rescue.

In the investigations and the enquiry which followed, the suitability of the Koepe winder for deep-level mining came under the spotlight and the controversy, mentioned earlier, about the suitability of the Koepe system for deep-level work was resuscitated.

They had nevertheless proved to be tremendous work-horses on numbers of South African mines. Bob Blair, Consulting Mechanical Engineer at Western Deeps at the time of their installation, had argued against their use, however. "I wash my hands of Koepe winders for this work," he said. He was overruled by the weight of counter opinion, however, and Western Deeps went ahead with the decision to install them.

Blair later developed a winding system of his own design, able to hoist greater tonnages from greater depths, and this system has been used extensively in modern shafts.

Barry Mudd, then Consulting Engineer, tells of other problems which that decision subsequently brought in its train. It was soon found that, used at such depths, the Koepe winder "eats rope". This proved costly in replacement as well as in working time lost during the rope changing operations. The problem seemed to stem from the fact that, with the immense lengths of steel ropes required over such depths, the long ropes twisted and were damaged.

Modifications were designed and implemented by Tony Gyngell, Consulting Mechanical and Electrical Engineer, who played an extremely significant role in the early design and development of the mine. He was assisted by a team of experienced engineers which included Des Bentley, Dick Davidson and Walter Wise. The modifications worked wonders and reduced the time for changing ropes from days to hours, but still did not wholly solve what appeared to be an almost insoluble problem. The manufacturers of the wire-ropes, Messrs Haggie and Sons, were consulted. In conjunction with South Africa's Council for Scientific and Industrial Research, the firm developed a special type of non-spin steel rope with an inner core that prevented twisting. This was the brainchild of Dr "Tiens" Koen, a research engineer at the CSIR, who later joined Anglo American.

In the Koepe system the wire ropes are not wound on to a headgear drum. The system utilises a continuous rope to which two counterbalancing cages are attached, so that as one cage goes down, the other comes up. Other types of winders handle cages individually and the ropes of each are wound on to a single huge drum at the head of the shaft.

In the case of Western Deep Levels, the continuous rope on the Koepe winder at Number 2 Shaft was 16 120 metres (9,7 miles) long and its weight, combined with the weight of skips, bridles and various attachments, amounted to about 234 tons, all of which, after this accident, lay in a tangled mass at the bottom of the shaft.

In his official report on the accident, Dick Lambert said: "... An analysis of events ... confirmed that the west skip struck the crash beams and deflection sheaves. The force of the impact broke the attachments and suspension gear, which were dragged over the drum, killing the driver and causing extensive damage to equipment in the hoist chamber. The ropes and attachments then fell down the east compartment, flailing about and seriously damaging the steel-work in the shaft ..." He suggests several possible causes, but in this particular accident the extent of the damage to the headgear made it difficult to pinpoint the actual cause.

The report concludes: "An investigation into winding accidents in Europe indicated that out of 60 accidents, 59 were attributable to either drivers' error or well-meant modification to the control circuits which had rendered the latter unreliable or unsafe."

The opinion against the use of Koepe winders in deep-level mines, initially offered in Germany against the weight of positive opinion in their favour, took on a particular validity.

Eric Brereton, who served at Western Deeps both as an Assistant Manager and, later, Manager, tells how, over several years, "the bugs were ironed out" in the Koepe system, and says the winders turned out to be "very successful work-horses". Those original four Koepe winders are still in use at Western Deeps and their high efficiency over many years endorses their success.

The damaged Number 2 Shaft eventually went into commission again on 5 January 1964, ten weeks after the accident: a costly delay, notwithstanding the fact that the resumption of operations was almost a month in advance of expectations. The cost of repairs together with the loss of profits due to curtailed production were estimated at about R1,66-million.

Despite this setback, Western Deeps registered a working profit of R7 382 000 for 1963.

The growth of operations was matched by the growth of the labour force. Indeed, personnel figures provide a handy barometer of the mine's development. At the end of that year, 1 227 white workers were on the mine's payroll, as well as 10 527 blacks. To handle the administration of the expanding work force and to assist with staff problems as they arose, a personnel department was created at the mine, headed by a manager and with a number of personnel officers and black assistants. Ben Petersen, formerly General Secretary of the Mine Workers' Union in Northern Rhodesia, was appointed Personnel Manager.

In 1965, in an attempt to level out the fluctuating availability of staff — a surplus in some years and shortages in others — an artisans' training centre was established in conjunction with a number of other mines that experienced similar problems. The training of apprentices began the following January.

Statistics quoted each year in the company's annual reports show impressive progress in all sectors of development and operations at Western Deeps. The figures rise in a bland upward curve, reaching higher and higher, always promising extensive future progress. But figures, impersonal and impassive, tend to hypnotise and lead one to overlook the extent of the human effort and energy which goes into their creation. Bald statistics breed no feeling for the labour and the sweat, the dangers and the pain, and the enormous courage and stamina required every working day in maintaining the thrust and momentum of progress. The historian Diodorus Siculus pointed out in the first century BC "that the production of gold is laborious," and no mining man — of any era — would dispute the continuing validity of that ancient thesis.

But if in those sixties the progress at Western Deeps was in many ways spectacular, it was not always smooth, nor was it all achieved without setbacks costly for the Company and heartbreaking for the miners. The overwind mishap at Number 2 Shaft was not the only accident in that decade.

Fire underground is a constant and frightening hazard on every mine. With the intensity of the ambient heat increasing with depth, the risk becomes greater as the shafts penetrate further into the rocky earth, and despite every precaution, Western Deeps was not without a number of underground fires in those early years. The longwall system of mining is particularly vulnerable since sections cannot easily be shut off and fires thus isolated.

In September 1964 the entire area at the bottom of the eastern longwall on 66 Level at Number 3 Shaft was sealed off to contain a fire which halted production until March the following year. In 1966 two further fires on the 94th and 100th levels of that shaft again brought local work to a standstill for several weeks, interfering severely with production on the Carbon Leader Reef. Notwithstanding these setbacks, however, overall production on the reef that year rose by 12 per cent.

Thankfully in neither case was there loss of life. In a most unusual occurrence on the surface in the following year, however, fifty miners died and many others were injured when, in a shift going on duty, some workers slipped and fell on the stairs in a subway. The press of workers behind them panicked and, in the confined space, trampled to death a number of their hapless colleagues. The official enquiry which ensued attached no blame to the mine management — nor was management to blame for deaths which occurred in two further accidents that year, one in which twelve were killed as a result of an explosion underground and the other when a rockburst caused the death of five workers.

The saga of accidents at that time created an unusually grim impression of Western Deep Levels. The deaths of fifty workers in the subway made the record look black indeed. But while those tragic deaths were on mine property, the accident, strictly speaking, was not brought about by any aspect of mining itself. It could have occurred at any large industrial plant or factory, or in a busy public thoroughfare.

On taking over as General Manager at Western Deeps in March 1966, Andrew Tennent said to all personnel: "... My first request is that you assist me in making this one of the safest mines in the industry. There has been a marked improvement in the accident rate in recent months, and it is my earnest hope that this will continue to improve ..."

His words were heeded. The improvement was made manifest when two years later — on 18 April 1968 — the Chamber of Mines awarded Western Deeps the "Millionaire Shield" for attaining one million consecutive fatality-free work shifts.

No miner needed reminding that that was certainly not a laurel not to be rested on! Danger casts shadows too long and too deep for those involved in mining ever to become careless or over-confident. Safety is a constant — not a sporadic — consideration.

The decade advanced. Reviewing the achievements of Western Deeps in 1968, the Chairman, Hermann Koch, told the Annual General Meeting during the following May that "... In view of conditions in adjacent mining areas it is considered prudent to increase the mine's water pumping capacity from twenty- to thirty-million gallons a day, and two-million rand will go towards this. Water actually pumped from the mine amounts to two-million gallons a day at present. . . ."

It was not long before the wisdom of that decision became apparent.

Towards the end of October 1969, just as the equipment for the increased pumping capacity of 20-million gallons had been installed, West Driefontein mine — adjacent — was suddenly faced with a major inrush of water underground and Western Deeps hastened to assist.

Co-operation was not — is not — limited to the region, however, nor, indeed, merely to South Africa. When Western Deeps made an urgent call to firms in Britain for the immediate supply of Newman-McEvoy control valves to be used in the emergency pumping operations, the suppliers in London reacted with alacrity and, within five hours of receiving the order, had loaded ten tons of equipment on to an aircraft at Gatwick Airport for immediate despatch to Johannesburg.

The flooding at West Driefontein was a major catastrophe by any standards. The magnitude of the disaster becomes clear with the realisation that the burst in a rock fault opened up a connection with the vast underground Bank Compartment in the overlying dolomite. This underground lake was estimated to extend over an area of some 60 square miles and to hold probably more than 100 000-million gallons of water. Moreover, the lake lay in rock above — above! — the levels at which the VCR was being mined and there were almost a thousand metres of water between the fissure and the surface of the subterranean lake.

Despite the frantic round-the-clock efforts and the ingenuity of engineers and workers, as well as the extensive pumping of millions and millions of gallons from the mine each day, for twenty-three critical and dangerous days the water defied all efforts to contain it and the life of the West Driefontein mine was itself in jeopardy.

Fred Bayley, the present General Manager at Western Deeps, was Underground Manager at Number 2 Shaft when this flooding occurred and he found himself responsible for carrying out the help promised to the neighbouring West Driefontein.

It was a harrowing time.

"We planned to develop a drill gallery in close proximity to the common boundary pillar," he recalls, "and from this position drill a number of large diameter holes into the flooded mine. Water was then to be drained through a system of high pressure valves and reticulated back to the Western Deep Levels pumps through a series of high pressure pipes. As an added precaution, concrete plugs were to be cast in all tunnels which were considered vulnerable to flooding."

Western Deeps could not allow the water to run wild through its shaft system. Its own mine could not be put in jeopardy. The risks were great. Frantically every possible counter-measure was set in train and carried on simultaneously.

"We broke every rule in the book," Fred Bayley says, "but there were no alternatives. This was a tough, emergency situation and time was of the essence."

Swiftly concrete plugs were designed and constructed to prevent any possible inrush of water. One plug was about a hundred feet long, to withstand the tremendous pressures. It was also feared that the high-pressure valves, which had to be hurriedly flown into the country, might not be strong enough to withstand the expected force and weight of the water. Fortunately these fears proved groundless.

Ted Wilson, the Assistant Manager at the time, and Fred Bayley were both present in the drill gallery at the time of the expected holing.

Tension rose as the diamond drill bit — almost four inches in diameter — approached the critical planned breakthrough point on the common boundary. Engineers at Western Deeps planned then to withdraw the drill rods in sections, uncoupling each in turn until the emergency holes were opened and pipes could be connected to take the flow of the water.

It is standard procedure in such operations.

But things did not work out that way.

"The intensity of the pressure was phenomenal," Fred Bayley relates. "As the steel drill rods encountered the water, they were fired like bullets from the casings and were buckled into what looked like spaghetti against the rock face of the drill gallery."

The operations nevertheless proved successful and Western Deeps was able significantly to assist in the dewatering process. "That was one of the most memorable experiences in my life," he says.

A Johannesburg newspaper, reporting the disastrous flooding at West Driefontein, said that "it would take over two years to reclaim the mine which accounts for eight per cent of the entire South African gold output."

Within a fortnight of the crisis, the well-known mining journalist, Paddy Cartwright, had written a gripping account of the whole episode, highlighting the extraordinary courage, loyalty and industriousness of all concerned. The tension throughout the account outstrips the excitement of much contemporary fiction.

The existence of vast quantities of underground water in gigantic caverns in the dolomite of the West Wits line and the threat to the mines of invading floods was used by the popular South African novelist, Wilbur Smith, in his thriller, *Gold Mine,* published a little more than a year after the flooding.

Indeed, Wilbur Smith, gathering material for his novel, spent several days at Western Deeps, talking with workers in all echelons who had played any part in the real drama.

The story was filmed soon afterwards — much of it on location at Western Deep Levels — with Roger Moore and the glamorous Susannah York in the leading roles and a strong supporting cast which included Simon Sabela, an outstanding black South African actor. It was released in South Africa in September 1974. *The Star* in Johannesburg rated the film highly, praising it not least for the portrayal of "the heroism of the men" involved in the quest for gold.

Underground water, however, is not the only natural problem facing mines. There are others, perhaps more dangerous and certainly more persistent. Of these, two in particular stand out. One is created by the intensifying rock pressures deep underground, the other derives from the ambient temperatures which go up and up as operations go down and down.

Apart from the acute physical discomfort and danger to life from heatstroke brought about by such high temperatures, the unnatural degree of prevailing heat gravely impairs work efficiency. For many reasons, a comfortable working environment is essential underground — as everywhere — and Western Deeps set about creating it. Early in 1965 at Number 3 Shaft refrigeration plants were installed and a further three 500 ice-ton units were added later. Similar cooling equipment was installed at Number 2 Shaft early the following year. It was estimated that, with the increased output from the Carbon Leader, it would not be long before further cooling equipment totalling 3 000 ice-tons of refrigeration would be needed at each shaft, while for the future that figure would be raised to some 9 000 ice-tons in each case. It is difficult to picture that quantity of ice. Imagine: a railway truck carries forty tons, so that 450 truck-loads of ice will then be required for cooling the two shafts! In 1968 alone more than R2,5-million rand was spent on refrigeration at the mine.

Two years later the Consulting Engineer reported that "Work is proceeding on the excavations for additional refrigeration plants to serve the workings below 100 Level. During the year 7 000 kW (2 000 ice-tons) of refrigeration was commissioned at each shaft, thus bringing the total installed capacity for the mine to 42 000 kW (12 000 ice-tons) of refrigeration."

With cooling becoming increasingly essential as mining went deeper, research continued to seek the best ways of creating effective working conditions underground.

It was Glen Griesel who master-minded the development of refrigeration at Western Deep Levels. He had joined Anglo in 1962 at a time when "Western Deep Levels was just taking off," as Theo Pretorius, Managing Director, said when presenting him in August 1987 with a long-service medal to mark a quarter-of-a-century's service with the Corporation.

At that stage the new mine presented "an enormous technical challenge with its tremendous depths, rock pressures and high temperatures." He went on: "Glen showed his mettle from an early age. He maintained, against all conventional wisdom, that the maximum amount of ice you could get down a shaft a day was not 3 000 but 12 000 ice-tons — and he was eventually proved correct. I think without that kind of technological input Western Deeps would not be what it is today." Glen Griesel was later put in charge of refrigeration matters for the entire Corporation and in that capacity "played a major role in transforming our deep, hot mines and creating good conditions."

Commenting on the extent of the refrigeration needs of Western Deeps, the *Rand Daily Mail*, in May 1966, pointed out that the refrigeration plant being installed there had "been designed and made in South Africa" by a firm in the Transvaal working in conjunction with engineers of Anglo American.

Progress in the sixties at Western Deeps would probably be most easily demonstrated by a comparison of various annual statistics of achievements at each end of the decade. But such comparisons at that stage would be neither meaningful nor realistic. Suffice it to say that by 1970 the mine had so developed that in that year Western Deeps milled more than three-million tons of rock and produced some 60 413 kilograms of gold as well as some 137 056 kilograms of uranium oxide.

The increase in staff may be regarded as a further barometer of progress. The white labour force had risen from 328 in 1960 to 1 556 ten years later, while over that period the numbers of black workers expanded more than five times — from 2 622 in 1960 to 14 720 at the commencement of the following decade.

Originally it was envisaged that Western Deep Levels would operate eight shaft-complexes, and the matter of expansion from the existing two shafts was kept constantly under review. From time to time the press speculated about the launching of a third shaft.

In his Chairman's report at the Annual General Meeting in May 1967, Hermann Koch said that "the question of an additional shaft in the eastern portion of the mine, with an additional reduction plant and thus a substantial increase in the milling rate of the mine, has been under consideration for some time." He indicated that a decision was dependent upon the results from Borehole U.D.14 then being drilled in conjunction with Western Ultra Deep Levels Limited and Witwatersrand Deep. Information obtained so far, he went on, "provides no clear justification for the very considerable capital outlay that would be involved . . ." It was felt that the best interests of the Company would be served by maintaining production from the existing Number 2 and Number 3 Shafts.

There, for the time being, the matter rested.

Following the abandonment of an international gold standard in the early 1930s, the United States dollar became the key currency in international trading and exchange. In the course of the 1960s, however, the weakening of the dollar led numbers of countries to exchange their dollar-holdings for gold. In an attempt to counteract the drain on gold held by the United States treasury, a two-tier exchange system was created in 1968. In terms of this the value of gold purchased by central banks was fixed at $35 an ounce while in the free market — consisting of non-governmental traders — gold was completely demonetised. A system of Special Drawing Rights created within the International Monetary Fund served as a reserve fund for central banks.

John Shilling, newly-appointed Chairman of Western Deeps, commented in his first annual report that these arrangements were of great importance to the gold mining industry and the country. He pointed out that the new arrangements announced in December 1969 by the International Monetary Fund, which affected the marketing of South Africa's gold, resolved uncertainties which had prevailed after the introduction of the two-tier system the previous year. Should the price on the free market fall below the officially established floor-price of $35 an ounce, South Africa would be entitled to sell its gold to the IMF at the official rate. A minimum selling price was thus established and security was ensured for the gold industry.

South Africa's Minister of Finance quickly pointed out that these new arrangements implied "that the key role of gold in the international monetary system is recognised and reaffirmed." Newly-mined gold could flow freely into the monetary reserves when there was no premium on the free market and, also, in certain circumstances even when there was a premium, he said.

The Chairman's review of 1969 concluded on an extremely optimistic note. He pointed out that recent market surveys had revealed how the use of gold outside the monetary field was growing rapidly and that "when speculative holdings now overhanging the unofficial market have been sold, demand may well equal, if not exceed, current production."

Increases in the average price of gold were slight in the early sixties, edging up a mere few cents a year: from R24,98 on ounce in 1962, for example, to R25,02 in 1963. In 1965 the price was R25,09 — "the same as for the previous year," the Directors reported. By these standards the upward surge in the closing years of the decade was a massive increase indeed, though it paled into relative insignificance compared with the soaring prices of the late seventies.

The acceleration of this trend, coinciding with the quickening of production at the mine, was of tremendous importance for the continuing development of Western Deeps.

The sixties, which had begun with the tragedy of Sharpeville, looked like ending on a high note, at least for the gold industry. For Western Deeps, now in full stride, the future looked bright — golden bright!

APPENDIX

On the occasion of the formal opening of Western Deep Levels mine in March 1962, the Public Relations Department of the Anglo American Corporation of South Africa, Limited, issued a brief pamphlet, "Transforming a blueprint into reality", acknowledging the "... outstanding achievement made possible only by the combined efforts of many people. ..." and in that note paid tribute to a number — but by no means all — of those people.

It read:

ACKNOWLEDGEMENTS

The opening of Western Deep Levels mine ahead of schedule, with a milling rate higher than that achieved by any other South African mine in its first month of production, is an outstanding achievement made possible only by the combined efforts of many people. Apart from those directly connected with the construction and development there have been a large number of people in the head office of Anglo American Corporation and the offices of the consulting civil engineers, the contractors and suppliers of equipment, and in other mining groups who have contributed to this project.

HEAD OFFICE: Consultants who were concerned with the mine in the early stages include Mr H.H. Taylor, now one of the managers of the Corporation resident in Rhodesia, Mr H. Clark, consulting engineer to the East Rand mines of the Anglo American Corporation Group, Mr H. MacConachie, consulting engineer to the Orange Free State mines of the Group, Mr C. Stent, the late Mr F.A. Unger, the late Dr J. Bancroft and the late Mr A.J. Brink. Those responsible for bringing the mine to production were Mr John Shilling, managing director, Mr J.B. Mudd and Mr G.H. Henderson, consulting mining engineers, Mr A.H. Gyngell and Mr D.M. Bentley, consulting mechanical engineers, Mr R.J. Adamson, consulting metallurgist, Mr J.H. Mortimer, assistant consulting metallurgist, Dr A.E. Waters and Dr H.C.M. Whiteside, consulting geologists, Mr K.S. Birch, the Group architect and his senior assistant, Mr A.C .Young, Mr F. Rodseth, Native Affairs consultant, Dr L.R.B. Bert, Mr A.J. Wasdell and his assistant, Mr H.E. Phillips, secretaries to the company, Mr A. Vice and Mr M. Ball, Group secretarial liaison officers, Dr G.F. Jacobs, personnel consultant (development), Mr J.H. Smith, Mr H.E. de Villiers, administration officer, Mr G.A. Hyatt, of the buying department. Mr W. Hedding and Mr R.L. Ingles, of the estates department, Mr H.T. Clausen and Mr W.H. Moyers, of the civil engineering department, Mr F.G. Abbott, chief draughtsman, and Messrs W. Willemse, P.H.W. Seaward, W.H. Burge, all of the engineering drawing office, Mr J.E. Porter, technical assistant and Mr T.E.C. Moore, of the estimating department.

MINING ENGINEERS: In addition to the consulting engineers of the Group, headed by Mr A. Royden Harrison, O.B.E., Western Deep Levels has received advice and assistance from mining, mechanical and ventilation engineers of other mining and finance groups, from the Council for Scientific and Industrial Research, from the University of the Witwatersrand, and the technical sub-committees of the Transvaal and Orange Free State Chamber of Mines. Acknowledgement is given to Mr F.G. Hill, technical manager and consulting engineer to Rand Mines Limited, and Mr C.J. Irving, consulting engineer to City Deep, Crown Mines and Blyvooruitzicht Mines, and to Mr R.R.M. Cousens, consulting engineer of the Gold Fields group, and to Mr L.T. Campbell-Pitt, previously consulting engineer and now a manager of Gold Fields of South Africa, to the general managers of City Deep, East Rand (Proprietary) Mines Limited and Crown Mines, and to other mine managers both inside and outside the Anglo American Corporation Group who have helped with suggestions and advice.

MINE PERSONNEL: Working with the mine manager, Mr F. Ludwig, the assistant manager, Mr R.N. Lambert, with the help of the works study and planning departments, was responsible for the planning of the mine layout and the essential details associated with the development programme. The progress achieved was due to the good planning and the outstanding leadership of the senior officials. Mr R.E. Davidson, the resident engineer and his chief assistants, Messrs A. Frolich, K.N. Hewson and A.L. Smith, carried the main burden of the construction work, the timeous completion of which played a major part in the overall achievement. Other officials who have helped to bring the mine to the production stage were Mr A.F. von Maltitz and Mr R.J. Parker, sectional engineers, and Messrs A.C.W. Comyn, W.C. Nienaber, D. Dawes and L.G.S. Wright, underground managers, who were responsible for the remarkable build-up in underground development operations. The master sinkers, Mr C.J. van der Westhuizen, Mr E.R.

Brune, carried out the shaft sinking. Other personnel who have played a notable part include the mine secretary, Mr B.E.H. Hulley, Mr F.D. W. Parry, the chief surveyor, Mr D. F. Rogans, the chief of planning, Mr A. Seelos, the senior model maker, Mr H. du Plessis, chief store-keeper, Mr H.L. Bowyer, mine accountant, Mr J.H. Taylor, compound manager, Mr J.H. Morris, sectional compound manager, Mr S.H.J. Taylor, reduction officer, Mr W.J. Ebertsohn, senior planning officer, Mr J. Thomson, surface surveyor, Mr P.J. Marais, chief sampler, Mr A.F. McCann, senior work study officer, Mr J.L. Beckwith, chief assayer, Mr R. Leuner, resident geologist.

OTHER ACKNOWLEDGEMENTS: Acknowledgement is also made to the company's consulting civil engineer, Mr C.R. Rigby, and to Mr D.A. Fordyce, the quantity surveyor, on whom the consulting engineers relied for all major civil engineering design work and for the certification of civil work carried out. Finally, the project has been brought to fruition because of the co-operation and excellent work performed by the many civil and mechanical contracting firms engaged on various aspects of the work, and the suppliers of major equipment and machinery.

[An apology was later extended to Mr Walter Wise, first Resident Engineer at Western Deep Levels, for the inadvertent omission of his name from the list of those who contributed to the early development of the mine. He was responsible for all the early construction work on site and at the time of the opening of Western Deeps was Assistant Consulting Mechanical Engineer at Head Office.]

CHAPTER EIGHT

". . . on the watch for anything precious . . ."

JOB: Chapter 28

Soon after production at Western Deeps came on stream and the published records of tons of ore hoisted each month became supplemented by details of the gold produced, increasingly frequent references to other minerals began also to appear both in the Board's minutes and the company's annual reports. These minerals were osmiridium, silver and uranium, which a bountiful nature had mixed with the gold, especially in the deeper Carbon Leader Reef. They were thus mined with the gold and, after separation, became by-products which were both useful and profitable in varying degrees.

In the heady 1950s, when the world was avidly seeking commercial sources of uranium for the development of nuclear energy, for peaceful as well as military purposes, South Africa, richly endowed with the mineral, seemed to have been dealt another valuable ace in the poker game of international trade. Furthermore, since uranium in South Africa was contained within the same ore body as the gold, its extraction after that of gold, against which the full cost of mining the ore would be chargeable, was likely to be less than that of uranium processed elsewhere. With production costs thus reduced — and profitability increased — the marketing prospects for South Africa's uranium were considerably enhanced.

For some years uranium had already been mined and processed at Blyvooruitzicht, on Western Deeps' northern boundary, and there was every reason to suppose that the presence of payable uranium continued into the reefs that extended southwards, particularly the deep-laid Carbon Leader. The assumption was confirmed by early assays, and Western Deeps' Prospectus, issued in January 1959, had announced that while the company was unlikely to become an approved uranium

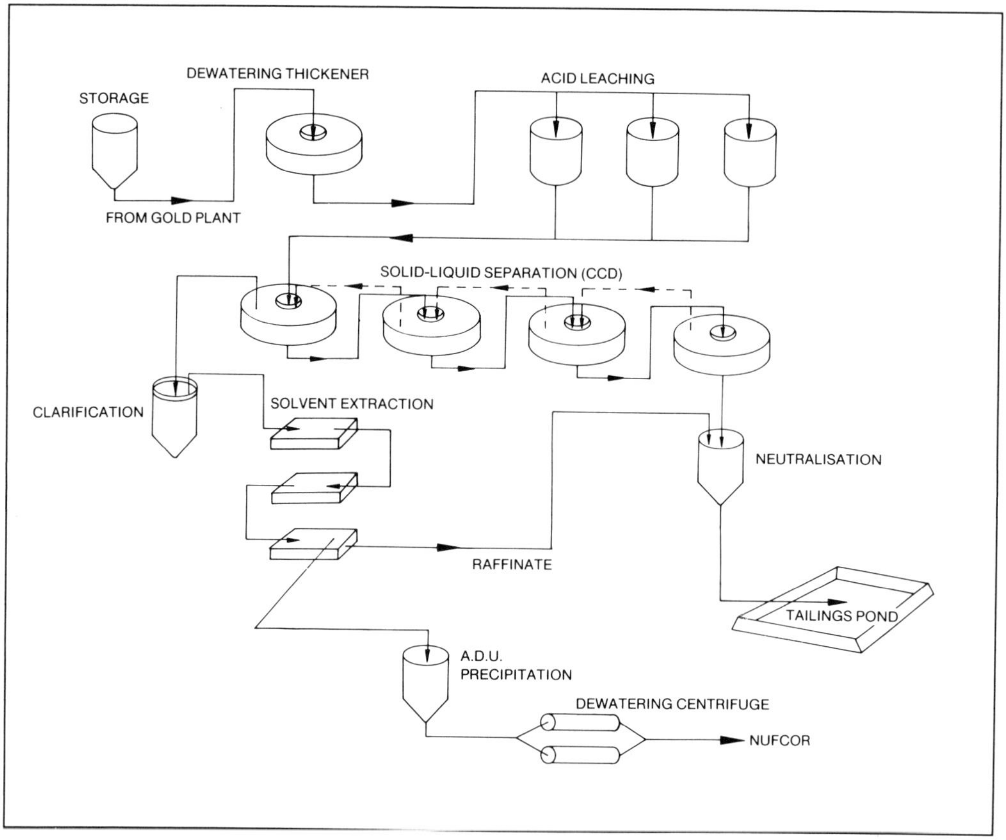

Uranium Plant: Process Flow Sheet.

producer at that time, "... the question of uranium production will be considered at a later stage in the light of the values encountered in development and of the then prevailing market conditions...."

Almost from its inception Western Deeps thus knew it had a second string to its golden bow: uranium could be a profitable secondary product once the mine was established. Clearly, however, that development had to await a more propitious moment — when the exploitation of the Carbon Leader had begun.

But if uranium was expected to become a substantial by-product at Western Deeps — to a degree, of course, then yet unknown — the prospects were less spectacular for the other by-products, silver and osmiridium, the latter being the name given collectively to a group of heavy metals: osmium, iridium, ruthenium and platinum. Quantities available were thought not to be significant.

The minutes of the Board meeting in September 1963 mention Western Deeps' osmiridium for the first time in recording that a one-year contract for its sale had been awarded to Engelhard Industries from March the following year. It was noted, blandly and without further reference, that this by-product had previously been sold to Messrs Derby and Company. The contract was renewed with Engelhard Industries for a further year in 1965. Twelve months later it was extended again. The contract soon ceased to be awarded on an annual basis and it was agreed instead that Western Deeps should sell its osmiridium to Engelhard's for the following fifteen years and, subject to three years' notice of termination, indefinitely thereafter.

Western Deep Levels produces very small quantities of this little publicised metal — "about a jam-jar full each month," Dr Frank Jantzon has said; "an ordinary household jam-jar. Very little indeed." That a mine which hoists millions of tons of ore annually should bother with the extraction of a mere "jam jar" of a by-product in itself indicates the value of that commodity: perhaps like a diamond of a few carats recovered from hundreds — thousands — of tons of ground. There has to be a good reason: simply it is there and is easily recovered at minimal extra cost. For the consumer, osmiridium is useful out of all proportion to the quantities mined and sold. Few civilised dwellings today are without its benefits. Apart from being used in the manufacture of scalpels and other surgical instruments, it is more widely used for filaments in electric light bulbs, millions and millions of which are everywhere taken for granted as darkness follows the setting of the sun around the world each evening.

As far as silver is concerned, considerably more than a jam-jar is mined at Western Deeps, though the quantities are similarly far from spectacular. The silver produced amounts to about 10 per cent by weight of the quantity of gold refined there. South Africa is not one of the world's major producers of silver, though international estimates establish the country as the fifth-largest source of the metal in the world today, with about six per cent of its total reserves.

At Western Deeps the ore, after being mined and crushed, is treated inititally at the mine to extract the gold. This is then sent to the Rand Refinery for final processing and purification. The waste rock is banished to the dumps which grow and grow, man-made hillocks that have become a feature of the surrounding landscape. It is at the Rand Refinery that the silver is separated from the gold, the amalgam from the initial process producing about nine per cent silver and 88 per cent gold. The remaining three per cent of the rough-refined metal consists of undefined impurities.

Recent years have seen a considerable upward trend in the price of silver and, too, in the amounts traded. From time to time, there have been rumours that world supplies will soon run out in the face of expanding use, and upsurges in demand have led to mini booms in sales. In December 1986 a United States financial journal, for example, predicted that the rising industrial demand would exhaust world reserves of silver by the year 1994.

By far the largest modern consumer of silver is photography, currently absorbing more than 40 per cent of the world's supply each year. The sensitivity to light of some of its compounds makes silver a basic component of photographic film, though by degrees scientists are now creating substitutes. A reasonably large proportion of the silver now used for photography is later recoverable, however, and doubtless — with supplies threatened — the future will see more attention being devoted to the recycling of silver from this source.

But if the production of silver is diminishing throughout the world, the opposite would seem to be the case in respect of uranium. Whereas, a short while ago, available quantities of uranium were unable to meet demands, now there is a glut, and hopes of a thriving secondary source of profit at Western Deeps have been dashed — at least for the present and probably also for the foreseeable future.

South Africa was established as one of the world's largest uranium producers well before Western Deeps got into its operating stride. By 1962 much of the necessary early research into the production of the country's uranium had been tackled and completed, so that by the time Western Deeps was ready to join other mines in producing it, most of the technical problems of extraction and refinement had been successfully ironed out.

Though the uses of uranium are a twentieth-century phenomenon, its discovery is not. That dates back almost two centuries to 1789 when the new-found element was named after the planet Uranus, itself discovered only a few years earlier. Its radioactive properties became revealed in experiments towards the end of the last century, while only in 1939 was nuclear fission discovered — by German scientists. For the world a new era had begun.

Though the presence in South Africa of radioactive substances had been known for almost thirty years, their significance was not really appreciated until the early war-time 1940s when the major Western powers began seeking sources of uranium for military projects. Then suddenly, with its military importance and its peace-time potential as a source of energy revealed, uranium assumed unheard-of value and universal importance.

Little can better illustrate the rapid rise in the significance of uranium than the fact — often quoted — that as late as 1901 Webster's Dictionary described uranium as "a worthless metal not found in the United States". The statement has been proved wrong on both scores!

While the sudden importance of uranium is generally associated with developments of the Second World War, and frantic searches for sources of supply were then initiated, it was long before — at the time of the First World War — that radioactive substances had first been detected in the gold ores of the Witwatersrand. At that time the discovery was not regarded as having any significant value, however. A few years after the war, an analysis of gold concentrates at a mine of the Central Mining Group had yielded a hitherto unknown mineral of the uranium family, uraninite, as well as small quantities of osmiridium. In 1923 R.A. Cooper, the South African metallurgist who had made this discovery, read and published a scientific paper on his findings.

The discovery prompted little or no significant reaction among either scientists or the public at the time. But Cooper's paper — which likewise then received scant attention — was later destined to have important consequences in the development of a uranium industry in South Africa.

In the 1920s and 1930s world demand for uranium was small, its uses being limited almost entirely to the paint and pigment and the glass and ceramic industries. Supplies came almost exclusively from Canada, Central Africa and Central Europe. The paucity of demand and the richness of those sources made it unnecessary to resort to others.

But World War II and the discovery of nuclear fission changed all that. Early in the 1940s scientists in the United States working on the top-secret "Manhattan Project" — the development of the atomic bomb — set about seeking and investigating new sources of uranium supplies. The matter was urgent and of the greatest importance. The search was given high priority. In the study of related literature available in the United States, Cooper's scientific paper, prepared in Johannesburg some twenty years before, was turned up. At once American scientists hurried to South Africa. The Prime Minister, General Jan Smuts, secretly requested to assist with supplies of uranium for the atomic bomb, turned to the gold mining industry. He asked Stowe McLean, then President of the Chamber of Mines, without delay to ascertain the extent of the uranium deposits on the Witwatersrand. The need for haste was accompanied by a compelling need for secrecy.

From the outset it had been realised that the Transvaal's auriferous ores were not overly rich in uranium. Indeed, by almost any standard their average uranium content was low. But the Witwatersrand had an enormous advantage over many other sources. Here the uranium was contained in vast tonnages of ore mined in the production of gold. With the basic cost of the mining already paid for by the gold, the cost of the uranium, a by-product, would be very considerably reduced. This low recovery cost was a key marketing factor. And the amounts of ore available in South Africa, however low-grade, were plentiful — estimated to be of the order of thousands of millions of tons!

The staggering estimates of quantities in the Transvaal led Dr Charles Davidson, at that time Chief Geologist of the Atomic Energy Division of Britain's Geological Survey, to say that, "Present evidence appears to indicate that the Rand may be one of the largest low-grade uranium fields in the world."

Promising assay results from batches of ore sent secretly to the United States early in the war led to the establishment of pilot uranium extraction plants at Blyvooruitzicht mine and two other mines in the Transvaal, supervised jointly by South African and American scientists.

The war over, the need for control in the mining and handling of such sensitive material led, in 1948, to the passing of the Atomic Energy Act by the South African Parliament; and then to the establishment of the Atomic Energy Board with wide powers to regulate all aspects of prospecting, production and disposal of radioactive materials, as well as the generation of atomic power.

In 1949, with the help and the approval of the South African Government, long-term contracts were entered into between the United States Atomic Energy Commission and the Transvaal Chamber of Mines for the supply of uranium oxide. Both the United States and Britain needed this basic material to build up their post-war atomic arsenals. South Africa was the answer to their prayers!

The Chamber of Mines played an extremely significant role in the establishment of South Africa's uranium industry, the importance of which is not always fully appreciated. Providing, as it does, a focal point of services to the mining industry, the Chamber was able quickly to bring together the engineers and disparate expertise necessary for the expeditious launching of this new South African industry on an efficient and cost-effective basis.

John Shilling, a director — and later Chairman — of Western Deeps, was one of those from the mining industry appointed to the Atomic Energy Board, thus ensuring and strengthening close co-operation between the industry and the authorities.

As the emphasis shifted from military to peaceful uses of uranium, in the early 1960s the Chamber set up the Uranium Sales Organisation, extending its involvement in production to marketing and assisting with the negotiation of terms and conditions of sale. It was not many years before the Organisation became the Nuclear Fuels Corporation of South Africa (NUFCOR), which not only acted as an agent for the sale of uranium from the mines, but also took over the central plant earlier set up for the calcining of ammonium diuranate slurry from the various extraction plants, converting it into uranium dioxide, better known as "yellow cake".

Further purchasing missions from the Combined Development Agency — the joint British-USA uranium buying organisation — visited South Africa in 1950 and again in 1951 to secure supplies of uranium for defence programmes. Contracts were signed for supplies for ten years from the commencement of production.

The contract price allowed for the full amortisation of the capital costs of the uranium extraction plants, totalling about £70-million sterling, as well as a fair margin of profit for the mines. By 1957 no less than ten South African mines were producing uranium, processed in seventeen plants. Production in South Africa reached its peak in 1959 when 26-million tons of gold ore, treated in those recovery plants, produced 6 400 tons of U3O8 as a by-product.

If, in the research which continued after the war, the emphasis was more on the application of atomic energy for military uses, there was no neglect of the development of nuclear power for peaceful uses. Research went on. With its supply potential accepted, South Africa became a partner in much of that nuclear research, and developed its uranium industry as a part of the co-operation with Britain, the United States and Canada, establishing its own uranium extraction industry and its own research facilities in less than ten years. In that research South Africa was no passive onlooker, but played an important part. Its scientists contributed usefully to the world's stockpile of nuclear knowledge. In due course this collaboration assisted South Africa's scientists in developing the country's own nuclear power projects.

Ewen ("Jumbo") Pinkney, for example, a senior metallurgist in the Anglo American Corporation, was assigned to research methods of extracting uranium from South Africa's low-grade ore. A patent, jointly with an American scientist, in 1949 crowned his endeavours and gave South Africa a commanding lead in this field.

For the mines the primary need was to find ways of recovering the maximum quantities of uranium from the hoisted ore with the mini-

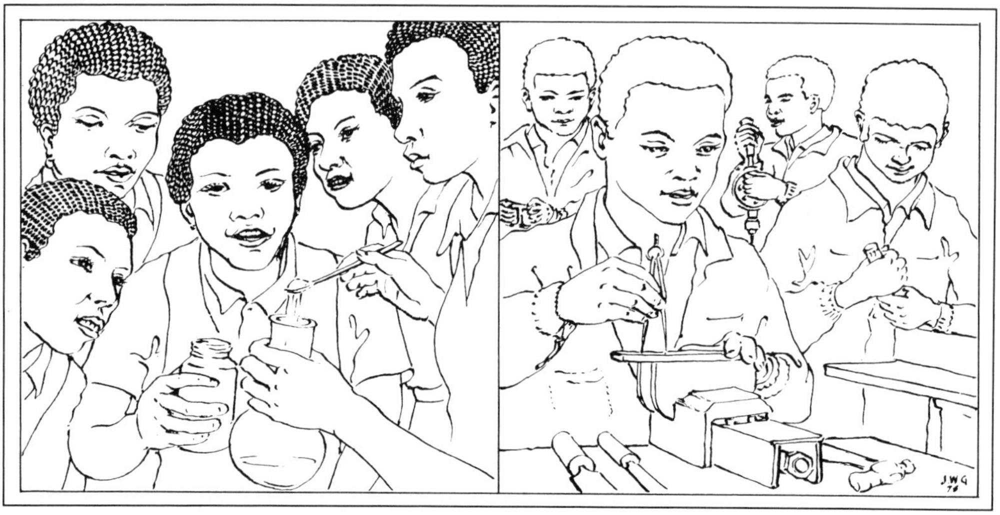

Senior students at Wedela Technical High School.

mum of disruption to the gold-extraction process. Research at the Government Metallurgical Laboratories assisted South Africa in making maximum use of the advantage nature had munificently bestowed on it.

The extraction of uranium involves a series of several highly complex processes. Any comprehensive description would take up an unduly large portion of this narrative. Perhaps a brief and simplistic outline will suffice.

First the host ore — in South Africa conglomerates and pyritic quartzites — is crushed to a fine powder. Where the uranium is a by-product of gold-mining, as in South Africa, this stage has generally already been completed in the extraction of the gold and the residues are processed for uranium. Mixed with water, the uranium compound (largely uranium dioxide) is then steam-heated and treated with various oxidising reagents to leach out the uranium. The resultant "pregnant solution" is then thickened and clarified by the removal of impurities before being subjected to one or other of a number of available chemical processes to concentrate the liquid slurry, purify it and then dry it to a powder — "yellow cake".

Western Deeps was too involved in the basics of its own development to be able, in the late 1950s and early 1960s, yet to expand its operations to the production of uranium and, anyway, uranium research still had a number of problems to solve, and many questions yet to answer. But the directors had not lost sight of this source of further revenue.

Suddenly, however, the market situation changed dramatically. Increased production of uranium in Canada and the United States had led by 1960 to world over-supply. The Combined Development Agency asked South Africa to place a ceiling on its local production to accommodate the decreases in current requirements and contract periods were accordingly extended to slow deliveries to agreed quantities.

The set-back should have taken none by surprise. In the late fifties world experts had been by no means unanimous in their views on the longer-term prospects for uranium. Writing in 1957, Sir Edwin (later Lord) Plowden, Chairman of Britain's Atomic Energy Authority, summed up the situation by describing what he saw to be three main lines of thought: "... the first was that the uranium industry was completely dependent upon military demand; that this would be satisfied within a few years; and therefore, within perhaps a decade, there would be a glut of uranium." His second argument, though admitting that the prospects for the civil uses of atomic energy were good, nevertheless took a pessimistic view of the future of the uranium mining industry. It was held that military demand would be satisfied before long and that soon the main use of uranium would be the initial fuelling of breeder reactors. But these reactors produced more fuel than they consumed; therefore, the civil demand, brisk though it might be while it lasted, would not last long.

Again, there would be a glut.

The third possibility was different. It was based on two factors: first, that military demand showed no sign of slackening; and, second, that many countries were hoping to meet their growing fuel requirements by using nuclear power stations of the types already in an advanced stage of development (i.e. non-breeder types). "In this event, the estimates of the future showed that not a glut but a shortage was likely ...", Plowden concluded.

It was common knowledge, he said, that "uranium is not a scarce material, but that it is as common in the earth's crust as tin and even more abundant than mercury ..." It was a comparatively plentiful material and occurred in over a hundred different minerals. Known reserves were enormous and, he predicted, more would undoubtedly be discovered with the passage of time.

Sir Edwin was, however, optimistic in his final conclusions. He felt that the need for uranium for civil purposes in the free world would increase, but that one of the conditions for the continuing high demand was "that it should be produced cheaply". This opinion held out considerable hopes for the success of the uranium industry in South Africa where production costs would be kept low by their dilution in the gold extraction process.

Plowden stressed, however, that though there were bound to be uncertainties of demand, nevertheless, "...the future of the uranium mining industry looks promising...."

With Western Deeps now producing satisfactorily and with a development momentum of its own, in the course of 1966 the directors were able to give consideration to the question of producing uranium. In his Report for that year, issued in the following May, the Chairman said: "... Some expansion in the demand for uranium became apparent in 1966, and the upward trend appears likely to continue. The stope value of uranium in the Carbon Leader Reef at this company's mine is 0,42 lb per ton, and a rise in the sale price of uranium might well present an opportunity for turning this uranium to profitable account."

At a Board meeting on 14 December the following year, the directors approved a proposal to erect a "reverse leach" uranium extraction plant at a cost of R4,5-million. Designed to treat 70 000 tons a month of upgraded slime arising from production on the Carbon Leader Reef, the plant was due to commence production within two years and then to produce 210 tons of uranium oxide annually. Shareholders were kept informed of the progress in construction and in May 1969 were advised that it was planned to commission the plant "early in 1970". In the meantime, the processing at the gold reduction plant was split into two streams and the high-grade slime deposited in a separate dam ready for treatment for its uranium content (see diagram).

Construction was indeed completed "early in 1970" — as promised — and by May the Chairman was able to announce that the production of uranium had commenced. The plant was located on the west side of Number 2 Shaft, conveniently close to the gold plant there. Building costs had been kept well within the estimated limits. To improve the operations of the solvent extraction plant and thus the grade of the uranium oxide produced, additions were made to the plant shortly afterwards.

The wisdom of placing the uranium plant close to the gold plant proved valuable when, in 1980, it was "decided to incorporate the leaching, filtration and clarification facilities ... into the existing gold plant." This resulted in the "saving of R1,8-million" on the original planned capital set aside for the extension of the plant.

Unhappily, about this time the demand for uranium slowed down sharply and prices dropped dramatically. Western Deeps therefore decided to stockpile its uranium products until such time — which the

Chairman optimistically felt would come soon — as production could be sold "at a satisfactory profit".

The anticipated upswing in the market was slow in coming, however, and it was not until 1972 that Western Deeps was able to show any profit from the sale of its uranium, and then only a paltry R7 000. The market continued to be lethargic and sales were small. The company's' profit and loss account for 1973 reflected a profit of only R6 000 from uranium sales, but suddenly the market improved and in 1974 the mine's profit on uranium sales soared to a dramatic R1,3-million.

In the next decade, however, those "uncertainties of demand" which Lord Plowden had mentioned, became increasingly manifest and the volatility of the uranium market made forward planning a headache. The future of the industry, which in 1957 was regarded with such optimism, did not fulfil the bright hopes which had been predicted for it. Sales were erratic, contracts were seen as elastic, and it was impossible to predict with any degree of certainty from one year to the next what quantities the market would, or could, absorb and at what price.

Despite the continuing vicissitudes of the market and yo-yoing prices, however, production of uranium oxide at Western Deeps not only continued, but trended generally upwards, as these figures show:

	Profit R '000	Tons Treated:	U-oxide produced: kg	Yield: kg/t
1970		544 206	137 056	0,25
1971		639 982	153 362	0,24
1972	7	616 626	159 068	0,26
1973	6	620 275	153 697	0,25
1974	1 328	616 000	136 248	0,22
1975	661	632 000	142 168	0,23
1976	1 719	603 000	140 001	0,23
1977	3 249	747 000	167 410	0,22
1978	2 341	686 000	183 365	0,21
1979	9 484	672 000	199 002	0,22
1980	3 501	162 900	213 000	0,13
1981	5 072	233 100	212 484	0,09

The picture was not always as simple, however, as the figures would indicate. In 1976, for example, the Chairman "expected that the company will have to purchase certain tonnages of uranium to meet its contractual commitments in 1978 and 1979. . . . the material can be made available from other producers within the Anglo American Corporation Group. Such transactions would have to be concluded at market-related prices which are likely to be in excess of our contracted selling prices . . ."

In the four years from 1973 the production of uranium oxide at Western Deeps had certainly not come up to expectations. This was due to three factors: grades had declined — as, indeed, had those of gold too; labour shortages had hindered milling, and several fires had interrupted production from the Carbon Leader reef from which all the uranium came.

In those same four years the price of uranium "rose more than sixfold", the chairman reported, until it stabilised in 1978.

In September that year the company announced its decision to extend the capacity of the uranium plant from the current 65 000 tons to 265 000 tons a month. The extension would allow 50 000 tons of reclaimed slimes to be treated as well as the total production from the Carbon Leader reef.

The cost of the extension was reckoned at R50-million and the enlarged plant was expected to produce an additional 175 tons of uranium a year when the plant was commissioned early in 1981.

It was not long, however, before the Chairman was having to tell shareholders that Western Deeps would not be able to sell its entire production this year, increased as it would be by initial production from the new plant extension. In that same year, 1979, he predicted that ". . . over-supply until the mid-1980s is likely to result in a continuing drop in price in real terms. . . ."

Glut and famine! It seemed difficult to get supply and demand to dovetail.

At Western Deeps the extension to the uranium plant was followed by a sudden unforeseen fall in world prices, not least as a consequence of the mishap at the Three Mile Island nuclear plant in 1979. For a while South Africa was cushioned from the full effects of a weakening market by the prevailing rand/dollar exchange rate and by the long-term nature of the contracts. But progressive weakening brought the market to a near standstill by 1983, and at Western Deeps it was decided to abandon uranium production and to convert the uranium treatment plant to handle gold. This was done — an unhappy conclusion to an interesting development. Grades had turned out to be lower than predicted. Of course, the plant could always be re-converted for the processing of uranium should the situation change in the future. Outstanding sales contracts were honoured by purchasing the required quantities of uranium oxide. This was done at better prices than Western Deeps could itself then produce it.

The move proved to be a fortunate one. In 1986 the disaster at the Chernobyl nuclear-powered plant in the USSR set the uranium industry back many years everywhere, inevitably with adverse consequences on world prices. Furthermore, the Chernobyl catastrophe was grist to the mills of the swelling anti-nuclear lobbies around the world. France and Britain, however, continue with their nuclear power programmes.

At the same time, the South African uranium industry suffered a further severe blow when the anti-South African feeling, prevalent in many countries, rose to fresh heights in the United States. In September 1986 there was a specific Congressional ban on uranium supplies from South Africa, and a US Supreme Court order banning all imports of uranium.

These decisions led to a protracted dispute — not, however, from any sympathy for South Africa. The ban was contested on the grounds that thousands of jobs in the United States would be placed in jeopardy by such a restriction and, moreover, that the consequent loss to America of revenue for the enriching of South Africa's uranium would amount to "hundreds of millions" of dollars. The protest was rejected and in July 1987 the US Treasury finally ruled that uranium ore and uranium oxide could no longer be imported from South Africa. Other markets remain available, however.

This was another generation of Americans: too young to remember how, forty-odd years before, South Africa had answered the United States' appeal for uranium in its national defence and in the vital defence of the West in time of war.

Nuclear power, nuclear weapons: all are emotive issues. Throughout the free world they spawn protesters and movements against the use of nuclear material. South Africa happens to be rich in the source material and thus must be regarded as a natural target for this sort of hostile action, as the past two decades have shown.

Compelled by these threats, South Africa has taken steps for its further development and survival. Increasingly isolated from the rest of the world by political forces, South Africa has turned to its own scientists who have developed methods of enriching uranium, while the Electricity Supply Commission has embarked upon a programme of nuclear power for the generation of the nation's electricity. The goverment has dedicated itself to ensuring that South Africa's nuclear resources will be used only for peaceful purposes.

But if the uranium industry did not fulfil earlier great expectations, it brought at least one other benefit to South Africa.

In his biography of Sir Ernest Oppenheimer published in 1962 Sir Theodore Gregory noted: "... attention has been given to the question of growing the guar plant in South Africa, the seeds of which, after processing, appear to offer a suitable flocculating medium for aiding filtration in the uranium plants ..."

Guar, cultivated largely in the East, is indeed used in the uranium extraction process in South Africa and quantities of this seed were imported from India.

Experiments, however, have proved the feasibility of growing this bean here. It is not only useful industrially, but also a source of highly nutritious food. This has provided some of South Africa's national states with a new and lucrative agricultural industry which, additionally, protects the uranium industry from the consequences of any boycott of this necessary industrial commodity.

The gold-mining industry has made — and is making — a major contribution to moves to increase South Africa's industrial self-sufficiency and further to implement the nation's independence of a hostile world in the field of power generation. In those efforts, Western Deeps has played its part.

With uranium now in plentiful supply within the country and without markets overseas, there is no longer a need for Western Deeps to produce it.

Without distinction the waste rock and processed residues are now hauled to the same dumps. From there, in due course, that waste will be put back underground again to refill the mined-out areas, unless, of course, the price of uranium soars to unprecedented levels and demand reaches unexpected and worthwhile heights.

But, apart from the gold, Western Deeps keeps on producing its "jam-jar" quantities of osmiridium and continues modestly to help stave off the threatening world shortage of silver.

CHAPTER NINE

Research: beating the problems

". . . Nature herself makes it clear that the production of gold is laborious . . ."

Diodorus Siculus (Roman historian: First century BC)

By the nature of his role in encountering situations hitherto unexplored, a pioneer seldom has precedents to guide him. His experiences may be advantageous or negative, but all bring fresh knowledge and his exploitation of them tests his ingenuity. From the research of new situations man's storehouse of knowledge becomes more richly stocked. In this sense, Western Deeps, probing greater depths, has contributed richly to the world's treasury of mining technology. When officially inaugurating Number 1 Shaft, Mr Gavin Relly said that since Western Deeps was pioneering mining at depth, "we naturally have had to find solutions to the problems which made mining at those extremes impossible in the past."

From the start, in its way, almost all mining on the Witwatersrand has been a pioneering endeavour for, as Professor Miklos Salamon has pointed out: ". . . There are no large mining fields anywhere in the world with geological conditions similar to those prevailing in the Witwatersrand Basin . . ." As a result, when gold was discovered there, technology appropriate to the exploitation of the Transvaal mines did not exist. It had to be created from scratch.

For Western Deeps, however, the situation has been somewhat different. It did not have to begin completely from scratch. One of a younger generation of gold mines, it became heir to a wealth of existing knowledge. But, being the first mine in the world to be designed especially for ultra deep level operations — deeper than any previously — it encountered, in its pioneering endeavours, a number of obstacles without precedent, even though their existence was often suspected and, indeed, often expected.

In practice, the elimination of new difficulties encountered in the mine has been tackled on two levels: one by the man-on-the-spot as

View of Number 2 Shaft: Western Deep Levels.

part of his continuing work, and the other more academically as designated research projects carried out in laboratories or on site, or both. In its particular way, each contributes to the solution of the problem, while co-operatively the fresh knowledge from both operations adds to the sum of experience and learning.

Whatever foreign politicians may say, scientists internationally acknowledge South Africa's leading position in mining technology. South Africa is also pre-eminent in the field of mining research. The range of that research is wide. It encompasses not only the study of a variety of problems underground, but also those on the mine's surface; it embraces not only engineering, geology and economics, but also the complexities of human behaviour in the multinational and multiracial work force unique to the country. Consideration is given also to the impact of machines and the working environment upon the psychology and physiology of the individual miner, who daily is faced by mental and physical stresses beyond those generally encountered by other categories of workers.

In a mining industry far less complex than today's operations, problems were previously tackled primarily when they arose. Now situations are specifically researched in an endeavour to anticipate problems and to find solutions in advance of needs.

Throughout the mining fraternity — both nationally and internationally — experience and knowledge are widely pooled. Mines tend to share what they learn, for the ultimate benefit of the industry as a whole. Furthermore, research in this field has taken on a new significance, as scientists and engineers of various disciplines collaborate in planned projects. For many years, too, the South African Chamber of Mines has maintained a Research Organization which has expanded to meet the manifold problems facing the country's mining industry.

The fruits of this local research are also shared freely with the rest of the world. At international conferences, scientists from all corners of the globe exchange knowledge and ideas, while technical papers and journals further disseminate the results of mining research.

Numbers of engineers from Anglo American and Western Deep Levels have participated in these conferences. In 1986, for example, at a mining conference in Singapore, engineers from Western Deeps described the major issues connected with mining at depth. Led by David Hamilton, Manager of Strategic Planning and Services, and Terry Hagan, Projects Head in the Research and Development Services, they read a paper which they had prepared with the help of Derek Lloyd and David Arnold, also of Western Deeps. At the same international gathering of mining and metallurgical experts, Van Zyl Brink presented an account of the very sophisticated means of attempting to predict rockbursts at Western Deeps, while colleagues Dick Solms and Alton Williams read a technical paper on shaft design and configuration in modern mechanised mining.

Terry Hagan was also a member of a team of researchers from Western Deep Levels which, in 1982, addressed an International Congress in Johannesburg on rockbursts and seismicity in mines. His confrères on that occasion were John Tanton, Mine Manager, and Trevor McCarthy, Mine Design Officer.

The century of mining on the Witwatersrand has seen a shift in the emphasis placed upon problems — both human and mechanical — specifically facing the gold industry. One of the main human problems in its early days was "miners' phthisis", caused by the intrusion of airborne dust — largely quartz dust — into workers' lungs. It was a scourge of frightening and devastating proportions in mining everywhere, the seriousness of which soon led to action by organisations in the Transvaal which tackled both cause and effect — prevention as well as cure. In the mines themselves this led to a greater awareness of a need for better ventilation and the elimination of dust underground. Minimum requirements in respect of dust control and ventilation in mines became entrenched in law, leading to rapid improvements in conditions and an accompanying drop in the death rate from phthisis.

As the Greek origins of the word indicate, phthisis is a wasting disease. The authorities have long been conscious of the scourge. As early as 1902 a commission was set up under Lord Milner to study means of limiting and eradicating the disease in South Africa's developing mining industry and research into eradicating pulmonary disease in mining has continued.

Through the years this disease of the lungs contracted by mine workers has been designated and defined in a variety of changing terms. Today it is known as "pneumoconiosis", more recently defined in South African legislation (1962) as a "permanent disease of the cardio-respiratory organs . . . which has been caused by the inhalation of mineral dust . . ."

Earlier — in legislation of 1912, 1919 and 1925 — the disease was generally referred to as "silicosis", while in 1946 the term "silicosis" was itself defined as "any form of pneumoconiosis due to the inhalation of mineral dust".

It has been necessary to entrench definitions in legislation to accommodate laws which, since 1912 in South Africa, have made mandatory the payment of compensation to miners who become victims of pulmonary disease, and legislation has been periodically amended through the years to keep pace with research and developments. South Africa has long led the world in the handling of pneumoconiosis, common to mining communities everywhere, and it was in South Africa, in 1911, that the first mass radiological study was made of silicosis. Miners everywhere have benefited from that research.

The importance of the malady was emphasised by the holding of the world's first international pneumoconiosis conference. Attended by delegates from many countries, it was held in Johannesburg in 1959.

In South Africa financial compensation paid to miners who become victims of pneumoconiosis is drawn from a government fund which is fed by a levy on all the mining companies. The amounts to be subscribed are determined, in consultation with the mine-owners, by an actuary appointed by the Minister of Mines. From the outset of its operations, Western Deeps has, of course, contributed to these funds.

Today pneumoconiosis has virtually been eliminated from South

Africa's gold mines: "engineered out of existence", it has been said, by water to lay the dust caused in operations underground. But other problems persist. Among them are some which, if anything, are more intransigent of solution than those created by dust, and which increase in severity as mining goes deeper. At present, the two main such problems are those of heat and of rock pressures, which without doubt are the most critical factors in the future of deep-level mining.

A paper presented jointly by Barry Mudd of Anglo American and Francis Hill of Rand Mines to an international Mining and Metallurgical Conference in Moscow in 1967 informed delegates that certain of the deep level mines in South Africa, besides being the deepest in the world, were also "amongst the hottest". They defined a deep mine as being one with a depth exceeding 2 300 metres or when the rock temperature exceeds 38 °C.

Ventilation is thus a key factor in operations, and from the start, in its planning and construction, Western Deeps followed the generally accepted pattern of twin vertical shafts at each complex, together with sub-vertical and tertiary shafts, one vertical being a downcast operating shaft and the other for upcast ventilating. The two are situated fairly close to one another. Huge fans supply the main volume of air which is also cooled when conditions underground demand it.

What temperature should the underground working area be? It must ensure a comfortable environment for miners as well as ensuring the maintenance of levels of productivity which drop sharply when the human body is subjected to excessive heat for any length of time. In addition, the risk of heat stroke must be eliminated from the work place. From a series of experiments by the Human Sciences Laboratory of the Chamber of Mines a standard "effective temperature" was worked out. That was fixed at 31 °C (87,8 °F) and for a while that was regarded as the working norm throughout the industry. All possible is now done to keep temperatures under 27,5 °C in underground working areas, however. At such a temperature the need for prolonged preparatory acclimatisation largely falls away.

The Laboratory has also given attention to the question of acclimatising workers throughout the industry before they actually encounter the temperatures of the work place.

The standard acclimatisation procedures formerly required recruits to exercise for predetermined periods over several days in temperatures which became progressively hotter. The intensity of the exercises was similarly gradually stepped up, while performances and body temperatures were carefully monitored.

This physical testing is done either in controlled conditions underground or in specially equipped gymnasia. Thousands of mine workers throughout the industry are acclimatised in this fashion. In South Africa in 1966 alone no less than 170 000 workers completed these preparatory acclimatisation courses. These procedures have proved effective through the years and although conditions underground are improving rapidly as equipment and knowledge become more sophisticated, the stage has not yet been reached when this physiological approach can be discarded. After all, as joggers and athletes will admit, exercise and training are essential for any work (or play) that is physical.

At Western Deeps, recruits, after a thorough medical examination, now undergo scientific physiological tests to determine their individual physical tolerance to heat conditions. On the basis of these tests, each worker is categorised and is allocated to a work environment commensurate with his particular ability to assimilate with temperature conditions.

Despite all efforts to reduce temperatures underground, there are inevitably working areas where it is impossible to keep temperatures below 27,5 °C.

Underground heat emanates from two main sources: from the surrounding rock and from the auto-compression of the air pumped from the surface. At Western Deeps some 32 per cent of the heat-load at deeper levels is due to the compression of the air itself. Calculations show that the temperature of air increases five degrees for every 1 000 metres of descent. The extent to which it heats up over almost 4 000 metres — down towards Western Deeps' lowest levels of operations — renders air alone ineffectual as a cooling medium.

Early in the next century, the deepest working place on Western Deeps will be at this depth. Here virgin rock temperatures of the order of 58 °C will be encountered, while the working levels expected to be reached by the year 2020 will be plagued by virgin rock temperatures in excess of 63 °C.

The combination of natural heat from the surrounding rock and from the compression of air pumped from above would thus create an environment in which sustained human labour would be impossible. And, to boggle the mind further, at Western Deeps plans are in train to mine deeper down, and, if possible, to tackle the tilted, dipping Carbon Leader at depths of 5 000 metres. Implementation of those hopes and plans depends not only upon the conquering of the physical conditions but, of course, also upon economic factors: whether income from the gold produced will exceed costs of its extraction.

Since ventilation is a key factor in these planned developments, considerable research is being done to develop the most effective and most economical systems.

From its inception Western Deep Levels has made provision for the cooling of the air supplied underground. Some of "the most powerful fans ever used in Southern Africa" were installed to ventilate the shaft systems. Made in South Africa, the exhaust fans were arranged in multiples to circulate some 800 cubic metres per second through each shaft system. Refrigeration plants were later installed both on the surface and underground to chill pumped air to ameliorate working conditions at depth.

The total amount of air circulated through all the shaft systems at Western Deeps is currently a staggering 2,3 tons per second.

At the new Number 1 Shaft, for example, the refrigeration system consists of a 42 megawatt plant on the surface. This will be supplemented by further refrigeration as operations develop, while ultimately, it is planned, 130 megawatts will be in use. The surface plant cools the air to 10 °C before it is pumped down to the working areas and also chills to 3 °C the water that is used to cool the rock at the working faces. This chilled water is fed underground by gravity flow.

Peak times of water usage underground coincide with peak times of demand for electricity. The chilled water is thus piped down at peak times and, servicing a hydro-electric plant, generates power to supplement the mine's electricity needs and to offset at least some of the costs of pumping the used water back to the surface.

Obviously ice has advantages over water as a cooling medium and studies continue into the economics of using ice as a coolant. Making the ice is costly, but the costs of pumping the ice are less than those demanded by the use of water since, compared with water, only about one-fifth the amount of ice is needed to achieve the same degree of effective refrigeration. At Western Deeps between 8 000 and 16 000 tons of ice would be required daily in any effective cooling system to maintain underground temperatures of less than 27 °C at depth. Investigations into the possible use of microscopic ice — as a compromise between ice and water, as it were — are now being pursued.

Clearly, if the deeper-lying gold is to be won, the savagely high temperatures which guard those reefs will have to be overcome. Intensive research into all aspects of ventilation and refrigeration is being master-minded by Derek Lloyd and John Cronje. Aspects of this current research into deep-level cooling present what is virtually a fresh field of science.

Economics dictate that every effort be made to minimise working costs — within the parameters of safety, of course. Power is an expensive commodity. At present, Western Deeps' monthly electricity bill runs to R5-million, which must make the mine a most valuable customer of the Electricity Supply Commission.

Indeed, at the close of 1987 Western Deeps was Escom's twelfth biggest client. Of the eleven bigger users ahead of it, five were municipalities, one an Anglo American gold mine, Vaal Reefs, while yet another top client was Anglo's Highveld Steel company.

Of all the problems facing modern gold miners, however, rockbursts are without doubt the greatest.

Somehow one does not imagine rock possessing the quality of sponginess in the slightest degree. The saying "as hard as a rock" creates an impression of absolute, unyielding firmness. But rock — with great compressive strengths — can store vast amounts of energy when under pressure. When the stress due to the weight of the overlying rock mass becomes too great, however, the rock fractures or slips along pre-existing faults, releasing the pent-up energy — often with violent effect and with considerable disruption of its environment.

The reduction and elimination of rockbursts presents the greatest challenge to deep-level mining. The bursts occur without warning; their coming is both swift and devastating and — for lack of warning — the consequences are often fatal. Rockbursts complicate not only the stoping operations in a mine but also create difficulties in other aspects of development underground.

David Ortlepp, Head of the Rock Mechanics Department of the gold division of Anglo American and an authority on rockbursts, maintains that at Western Deeps "the loss of production due to rockbursts amounts to approximately five per cent — which is of almost negligible importance." He goes on to say: "Thus the seriousness of the rockburst problem lies entirely in its threat of death and disablement and its effect on worker morale. . . ."

Just as research has eradicated a number of other mining problems, so intensive research continues into ways of reducing and eliminating rockbursts, examining what causes them and looking at the ways in which their approach can be detected, as well as seeking methods of protecting life and limb from their ravages. Rockbursts raise infinitely more problems than both dust-induced disease and heat and are also more dangerous.

Much has been done in recent decades to combat the threat of rockbursts. When the long-walling system of stoping was first introduced into South African mining by Francis Hill in 1941, it was seen as a major step forward in the reduction of rockbursts. Western Deeps initially adopted this system in its stoping but later, as mining spans and the rockburst hazard increased, modified the lay-out to a mini-longwall method by leaving systematic stabilising pillars. In due course, risks will be further reduced by supplementing the stabilising pillars with a backfill of de-slimed mill residue.

Of course, leaving large stabilising pillars and thus not mining areas of the narrow reef — as necessity is now seen to demand at depth — reduces the ore yield by the same percentage as the size of the untouched reef is in relation to the whole. The matter thus becomes complicated by the introduction of basic economics: how little need be left untouched without endangering safety? It is a fine point, but always the answer to the equation tilts prudently towards safety. That is paramount.

Conditions at Western Deeps make it necessary to leave 15 per cent of the reef unmined for the creation of the essential stabilising pillars in the longwall operations.

While their incidence and severity increase, rockbursts are not only a phenomenon of recent decades. As far back as 1908 an official commission was set up in the Transvaal to investigate the problem. Other official enquiries have followed at intervals.

At that time South Africa's gold mining industry was of course only a fraction of the size it is today. Records show that the annual casualty rate in the country's mines in 1906 was five per thousand workers, of which 20 per cent was due to rockfall or rockburst accidents.

Algie von Holdt, Consulting Engineer to Western Deeps, reported to the Directors in 1976 on the continuing "long-term programme which is being carried out in collaboration with the Chamber of Mines Research Laboratories . . . to investigate the causes and effects of seismic activities . . ." It was a project initiated in 1973 by Anglo American and other mining houses in conjunction with the Chamber of Mines as part of an intensified programme throughout the industry to improve productivity and safety.

From data collected by underground instruments, von Holdt said, statistics were "being compiled from which it will be possible to evaluate the effects of longwall shape, geological disturbances, stoping width, hydraulic props, rigid forms of support, waste stowing and other safety measures. . . ."

For management — and, indeed, for all concerned — rockbursts were a headache that would not go away. A major portion of the

Chairman's report for 1979 was devoted to the matter. "The major unsolved problem in deep-level mining below 3 000 metres is the incidence of rockbursts," said Gerald Langton. "To a certain extent rockbursts have been limited by longwall mining which minimises the formation of remnants, by good mining geometry and by the use of rapid-yielding hydraulic props. Despite these measures, however, the incidence of rockbursts is still far too frequent and when these occur during the working shift they usually result in injuries and often in loss of life. Besides the human tragedy associated with these events they create an unfavourable image of the mine which in turn leads to recruiting problems.

"Naturally, the majority of bursts occur after the blast when the stope faces are not manned and consequently do not result in casualties. However, these bursts normally cause severe damage to the workplaces which takes time to repair and results in loss of production. Furthermore, they have a demotivating effect upon the miner and his team. It is estimated that production lost in 1979 amounted to 44 000

Modern hydraulic mine props.

square metres and as the depth of mining increases the rockburst hazard and losses could become even greater."

Langton went on: "Where stabilisation is provided by a systematic pattern of strike pillars, it has been demonstrated that an 85 per cent extraction rate will result in a 50 per cent reduction in the number of rockbursts. . . ."

Current research looks not only into ways of eliminating rockbursts but also — until their hazard has been eliminated — into ways of detecting the threat before the burst occurs. The reduction and eventual elimination of the rockburst hazard is being tackled as a matter of the highest priority. Depth would seem to be one factor in the frequency of rockbursts. It is thus not surprising that, of all South African gold mines, Western Deeps faces greater hazards in this regard than most other mines.

Two thousand years seem to have done nothing to cast doubts upon the perception of the Roman historian, Diodorus Siculus, that "Nature herself makes it clear that the production of gold is laborious . . .". In 1986 the Chamber of Mines echoed the sentiment, saying, ". . . The statistics show simply that underground mining is hazardous. The Chamber of Mines is well aware of this. . . ." The note went on to point out that the Chamber and the mines themselves devote ". . . enormous time, effort . . . and hundreds of millions of rands in trying to reduce the risks of accidents and fatalities. . . ."

The hazards are not limited to South Africa's mines. Mines elsewhere face similar problems. Experts from overseas have said firmly that this country's mines are no less safe than their American and European counterparts, and a senior official of the US Mine Safety and Health Administration has pointed out that ". . . South Africa has some of the most advanced safety-conscious equipment and safety training we have seen."

Western Deeps maintains a large staff of specialist scientists in its Rock Mechanics Research Laboratory, which is equipped with some of the most up-to-date monitoring machinery in the world. To enter that laboratory at the mine is to visit another world. Doors fit tightly, windows are sealed: the temperature within is controlled and constant. There is no dust. Cleanliness is clinical. Nothing is allowed to upset the working of the rows of electronic monitoring equipment that hums and blinks throughout the days and nights. Lives depend on its efficiency and upon the expertise of the scientists who control it.

Through the large windows of the office block one sees in the hot sunlight the indigenous trees dancing noiselessly in the wind, casting on the panes silent shadows of a world whose physical isolation is breached by electronics rather than by human contact.

Terry Hagan, Head of Rockburst Research at Western Deeps, talks about the problems and tells of efforts to solve them. "Without doubt," he says, "the rockburst problem is the most serious we have to face in mining."

It becomes clear, as the talks proceed, that the officials of the research team are well aware of the gravity of the tasks which face them. They confront the challenge with confidence and responsibility.

At present, back-filling is not an alternative to partial extraction. Rather, one supplements the other in endeavours to reduce the energy available for seismic activity.

The system of back-filling is not new. It has been used on the Witwatersrand mines for almost a hundred years. The practice involves refilling mined-out areas with rock waste or with residue sands from the mills, or both, instead of piling it on to mine dumps. But, though the process is not without problems — and therefore additional costs — it is believed to be an effective aid in solving the rockburst problem. Furthermore, back-filling of mined-out areas holds at bay the heat radiated there, thus helping to reduce temperatures at the work-face. In 1987 a capital project was authorised "to establish classified back-fill plants at the three shafts" at Western Deeps. The installation of a pipe reticulation system will enable fill to be placed in most of the critical areas underground by the end of 1989.

Conditions at Western Deeps make it necessary to leave 15 per cent of the reef unmined, for the creation of the essential stabilising pillars in the longwall operations. Consequently only 85 per cent of the reef's potential is being tapped.

It has been pointed out, however, that "historically 5 per cent of the reef has been left as random pillars because of mining difficulties related to depth. It is expected that these problems will be overcome in most cases with pillar mining. Thus stabilising pillars will result in 10 per cent of reef being left unmined compared with previous practice."

This will lead to some reduction in the operational life of Shafts 2 and 3 at Western Deeps. In 1982 the life of those shafts was considered to be another thirty years. As a consequence of pillar mining, this will be reduced to twenty-seven years. The amount of gold left untouched there is estimated to amount to some 140 000 kilograms. Relative to production and after-tax profits, the impact of that on present net value is considered minimal.

Sitting in the isolated quiet of the research laboratory at Western

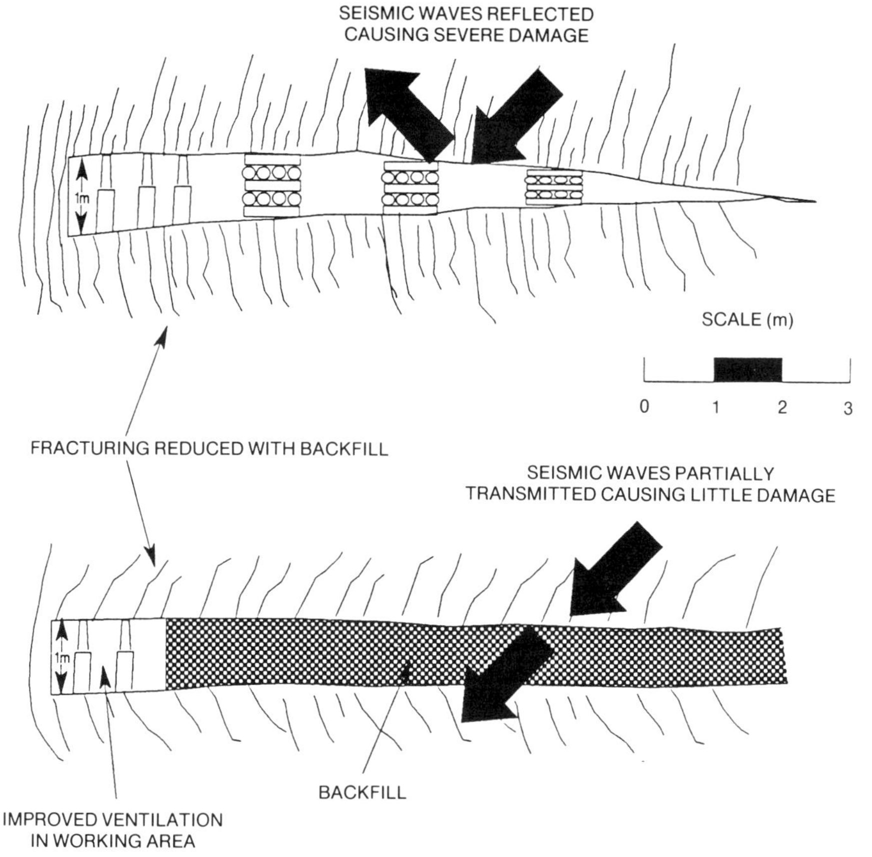

Section of two underground excavations, one of which is backfilled with residual sands. Advantages of backfill include improved roof conditions, reduced seismic damage and better ventilation.

Deeps, Hagan goes on to tell of the tremendous force of rockbursts, comparing them to natural earthquakes.

"The energy released in a major seismic event is equivalent to a minor nuclear explosion," he says. "It registers around the world as a minor earthquake. Some rockbursts in the Klerksdorp area have registered the equivalent of five on the Richter Scale." Some of these events have prompted interpretations — possibly characterised more by political than scientific discrimination — that they were indeed due to underground nuclear testing.

Backfill in stope.

His colleague, Van Zyl Brink, Head of the Prediction Project at the Laboratory, tells of the work being done on the prediction of rockbursts and of the attempts to develop an effective early-warning system.

He explains the similarity between an earthquake and the seismic event which sometimes causes rockbursts. They are basically the same phenomenon, one created by nature, the other man-made. Both are seen as failures of geological features usually brought about by changes in stress conditions. Natural changes are taking place all the time all over the world — caused, for example, by the constant moving of the earth's tectonic plates.

The similarity between nature's earthquakes and mining's rockbursts leads to the use of the same sort of equipment for their detection, though recent developments have led to the making of more specialised instruments unique to mining research. Their heightened sensitivity will prove useful also in earthquake detection and prediction. "Deep mining is now a laboratory for some seismic research," says Van Zyl Brink.

As experience grows, the accuracy of these predictions increases. The mine's early-warning system of seismic detectors scattered systematically about the mine and linked to computers in the research laboratory will hopefully one day make it possible to predict threats with greater precision and, when necessary, evaluate with refined accuracy the possible need to evacuate workers from threatened areas.

Dr Dennis Etheredge, Chairman of Western Deeps from 1974 to 1977, emphasised the value of advance warning of a threatened rockburst. "Even as little as three minutes," he said, "helps to get the men to places of safety." The research laboratory hopes to do better than that.

Not only does a work-stoppage — for whatever reason — cost money, but in the circumstances of danger it has a tremendously adverse influence on the workers' morale. False alarms breed a dangerous cynicism about the efficiency of the alarm-prediction system. Economics provides only one reason why it is necessary to eliminate false alarms. There are others too. In the past the underground managers shouldered the responsibility for halting all work in the event of any threat at the work-face. With only local circumstantial evidence, prediction became a matter of experience and was essentially subjective. With the development of new techniques, however, and increasingly sophisticated and sensitive equipment, such decisions become objective and much of that responsibility now shifts from the underground manager to the laboratory technician, creating new dimensions for management.

Van Zyl Brink believes that shortly — "within the next five years" — a practical prediction system will exist, but admits that "false alarm and no-alarm situations will still occur." The system will thus not be absolutely reliable "and subjective decisions will have to be made by the seismologist and the mine management".

Experience so far leads to the conclusion that in future the prediction of rockbursts will be made in stages. "... First a long term warning will be given (say, seven to fourteen days before the event) with about 30 per cent reliability." He goes on: "This will be followed by an intermediate warning (say, one to seven days before the event) with about 50 per cent reliability. Finally an imminent warning will be given (say, ten to eight hours before the event) with about 70 per cent reliability. ..."

Western Deeps once had the reputation of being an unsafe mine, and indeed for a while it had a poor fatality record compared with other

Transvaal gold mines. With the extent of the current research, however, the picture has changed substantially.

Writing in the *Journal of the South African Institute of Mining and Metallurgy* in April 1980, Anglo American's Rulan Heunis examined causes of fatal accidents at Western Deeps — "a typical deep level mine" — over a period of three years in the late 1970s. ". . . The combined fatality rate due to rockbursts and rock-falls," he said, "was 1,75 per thousand employees per annum, or 80 per cent of the total fatalities for the mine. The rockburst contribution equalled 1,0 per thousand per annum, or 46 per cent of the total."

He compares those figures with statistics for five typical shallow and medium-depth gold mines. The comparison makes it clear that, as the average depth of mining increases in the future, "rockbursts and rock falls will tend to dominate increasingly as the major cause of fatal accidents, the relative importance shifting from rock-fall to rockburst accidents. . . ."

Comparing these rates with the "corresponding national average mortality rate of around 0,8 per thousand per annum for males aged 30 to 40, it is clear that, in deep-level mines, the additional risk posed by rockbursts and rock falls is indeed significant. . . ."

David Arnold, Manager of Rock Mechanics in Anglo American's Western Division, talks about deep-level mining of the future, explaining that problems will surely differ with increasing depth. It is expected that support problems will increase as mining goes deeper. It is thought that a stage will be reached at which tunnelling will be critically and adversely affected by overhead stresses, unless adequately protected. He defines "deep" as mining below 3 000 metres. Western Deeps is already studying the feasibility of such operations. He is quick to add, however, that there is no final definition of "deep". Mining techniques will be the final arbiter.

Further explaining the functions of rock mechanics research at Western Deeps, Arnold outlines the broad spectrum of activity — in the designing of shafts, for example, the configurations of mining, the configuration of tunnels, and so on. Modern computer techniques have facilitated and speeded up the work, even to the extent now of indicating how sections and blocks underground should best be mined, supported and protected. These techniques are becoming increasingly valuable in the designing of mines.

In all this work there is close domestic liaison between the research and service divisions at Western Deeps as well as with other mines and, too, with the research laboratories of the Chamber of Mines, which has the biggest rock mechanics research laboratory in the world. "Need has bred the size," Arnold says.

The Rock Mechanics Department at Western Deeps has also grown in size and stature in recent years. In January 1966 the *Deep Levels News* reported: "... Tucked away in four offices in the West Wing of the Central Offices at Western Deep Levels are Messrs Frank Jantzon and John Cook and Dr Dieter Hallbauer — the backroom boffins of our Rock Mechanics Department...." But if the department has grown greatly since then, its functions and the dedication of its workers have remained unchanged. That same issue of the *Deep Levels News* went on to say: "... The science of rock mechanics is playing its own extraordinarily effective role in improving our mining methods and, of equal importance, the department's research is helping to make mining conditions even safer ..."

Progress in rockburst research has been significant, though there is no sense of complacency. Men and machines collaborate closely in a strange harmony, each contributing to the development of the other in the pursuit of common goals.

Something of this progress is manifest in the annual report for 1987 of the Consulting Engineer, Fred Bayley. He points out that in the year "... The number of rockbursts showed a welcome decrease. A 15 per cent decrease in seismic events, having a magnitude of 0,8 or more on the Richter Scale, was experienced. This is attributable mainly to the 40-metre-wide stabilizing pillars that have now been introduced...."

Andrew Tennent, General Manager in the latter half of the 1960s, recalls how Western Deeps was one of the first mines to install seismographs in an effort to determine the location of underground rockbursts. "The system was primitive compared with today," he says, and relates how "three pieces of string" were used in conjunction with the strategically placed seismographs to establish, by triangulation, the area of any rockburst which registered on the instruments.

It is a gigantic leap forward from those "pieces of string" of twenty years ago to today's sophisticated computers. Frank Jantzon points out, however, that the string model analogue of the sixties — then state-of-the-art equipment — "worked exactly on the same principles which nowadays are simulated into computer programmes."

The developments from the simple procedures of earlier years to the present sophisticated technology demonstrate not only the progress of research in this vital field, but emphasise also the continuous and continuing concern in efforts to ensure maximum safety in mining at Western Deep Levels.

CHAPTER TEN

People

All over the world "apartheid" has become a shorthand description of South Africa's way of life. Overseas the word has an infinite number of interpretations, none of them benevolent, few of them accurate. Foreigners tend to believe that all white South Africans subscribe to its practice.

Anglo American has long led a campaign for the amelioration of race relations and the removal of discriminatory legislation affecting the heterogeneous population of the country. In this drive to end racial discrimination, Anglo is joined by most of the country's larger business houses and industrial concerns. Often their liberalism has clashed with the conservatism of the country's political rulers. By degrees, however, with the advancing education of the blacks, the tide is turning and the policy of apartheid, cracked and eroded, is now disintegrating. Throughout the world the historic stigma will doubtless long remain, however. The brand has burned deep into the country's hide. For those with designs upon South Africa, the stigma is too valuable a weapon to be allowed to vanish.

As an integral part of the Anglo American group, Western Deep Levels has always nurtured this sense of social responsibility among its workers and, within the parameters permitted by law, has striven for harmonious industrial relations and for social justice in all its operations. From its inception, Western Deeps absorbed the philosophy and the practice into its own scheme of things. To have done less would have been unthinkable.

It was Anglo's social philosophies and practices which were emphasised by the Public Orator at Oxford University, when, in October 1952, a degree of Doctor of Civil Law (*honoris causa*) was conferred upon Sir Ernest Oppenheimer. After speaking of his considerable political

Swimming pool at Wedela Club, overlooking part of the town.

and economic achievements, the Orator went on to say: "It is more important to mention his humane spirit in providing for the health and general welfare of miners . . ."

This spirit is manifest in the housing of staff and workers at Anglo American's mines, perhaps most of all now at Western Deeps.

Ten years before the commencement of operations at Western Deep Levels, Sir Ernest, launching Number 2 Shaft at Welkom Gold Mine in May 1947, said that it was Anglo's aim to have the best possible housing conditions for white staff and materially to improve the conditions in the housing of black workers. He felt Anglo's policy in this regard ought to be "to create, within a reasonable time, modern native villages" that would attract workers from all over South Africa.

The concept did not take root, however. A new government came to power a year later and, dedicated to its policy of apartheid, frowned upon the idea, insisting that workers should live in their own territories and come to white areas only as migrant labourers.

On 12 June 1952, Mr Harry Oppenheimer, Member of Parliament for Kimberley City — a seat formerly held by his father — rose in the House of Assembly in Cape Town to address the Minister of Native Affairs, as the Cabinet portfolio was then known. "I would like to ask the Honourable the Minister," he said, "whether he would be kind enough to define his attitude in regard to the experiment which is being made on certain of the new gold mines in the Free State, by the establishment of villages on the mines for married Native employees. What is being done is not entirely new, because the gold mines, as the

Honourable Minister knows, have for a long time had a certain number of married Native employees who have been housed with their wives on the properties. But those detribalised Native employees on the mines were few in number and were almost entirely limited to Native clerks working in the offices of the mines on the surface.

"Now an experiment has been started on certain mines in the Orange Free State to see whether married accommodation could not with advantage be provided for other classes of Natives, including Natives who work underground. . . ."

The Member for Kimberley City was at pains to point out that "this scheme has no ideological basis. It is simply a scheme which is intended to meet the present conditions and the difficulties with which the mines are faced. . . ."

At that time South Africa's mines were dependent in large measure upon labour from territories beyond the country's borders. It was hoped that this would be a temporary situation and that, with time, more and more South African blacks would come to earn their livings on the gold mines. ". . . I should have thought," Mr Oppenheimer went on, "that if these people are to be taken off the land and . . . if their source of livelihood is going to be the mining industry in the so-called European areas, then they should be established in towns not in the reserves, but near the industries where they are going to be employed. . . ."

His speech was cut short by the clock just as he was saying "I think there is a fundamental confusion in the hon. Minister's thought in this matter. . ."

The Minister of Native Affairs, Senator Hendrik Verwoerd, later to become South Africa's Prime Minister, replied the following day. His answer was firm and left no room for doubt.

On the question discussed by the Honourable Member for Kimberley (City), Mr Oppenheimer, the Minister said he was opposed to any development of married quarters at the mines. He went on: ". . . my Department has been instructed to investigate the position very thoroughly and to stop the development of such villages as far as possible. Let me give my reasons for this. Within that Free State gold-mining area every mine can then establish its own Native town with married quarters. That will then mean a series of Native towns. . ." The Minister suggested that these would eventually become "black spots spread throughout the whole Free State mining area" which, when mining ceased there, would continue there.

He conceded that "the mines need a certain number of experienced married Natives such as boss boys but our view is that there should be married quarters only for those who are needed on the mines for night duties or for emergency duties . . ."

Undeterred, Anglo pursued the matter outside Parliament by correspondence with the Department of Native Affairs which, some fourteen months later, "as a special concession" agreed that "Native personnel, who are employed on a permanent basis because of specialised training and experience and who are required to be continuously on the mine property, may be accommodated in a limited number of married quarters on each mine." The concession did not apply, however to "foreign Natives".

Through the years Anglo did not lose sight of this matter of housing for black mine workers and Western Deeps — new to the mining scene since Dr Verwoerd's ruling in those early fifties — repeatedly added its weight to the continuing representations for improved amenities and better housing.

In his report for 1975, the Chairman, Dennis Etheredge, said: "In the longer run our goal must be to stabilise the industry's black labour force as far as possible . . ." The provision of more married housing would be of inestimable help in that regard," he said, adding, "It is most encouraging that, following discussions between the industry and Government, we have been advised of new arrangements which will permit senior married black employees from South Africa and the homelands to be housed with their families in black residential areas where they are reasonably close to the mines."

No sooner said, than done! In the following year the Chairman reported that "Western Deep Levels has received permission from the authorities to erect, in conjunction with Elandsrand Gold Mining Company Limited, a new village for senior married black employees. The first phase of the project will comprise 600 three-bedroomed houses . . ."

In December 1978, more than a quarter of a century after his plea in Parliament, Mr Harry Oppenheimer officially opened Wedela township, a residential area for senior black employees and their families. It lies between Western Deeps and Elandsrand, two Anglo mines, from which its name has been fabricated.

Fred Bayley, then Consulting Engineer to Western Deeps, pointed out in his report for 1978 that "one hundred and twenty-six houses have been completed, of which 26 are now occupied, and a further 238 will be built in 1979."

The town has grown since then. Now it has fine brick and tiled houses clustered around small precinct parks, and the surrounding low rolling hills add beauty and character to its appearance. While the designs of the houses differ, all are roughly the same size, comprising a lounge-dining room, three bedrooms, kitchen, bathroom and toilet, an enclosed courtyard and a carport. Rentals range between R5 and R10,75 a month. Water and light are subsidised. Maintenance is the responsibility of Wedela's housing department, which also employs a horticulturist to assist residents in establishing gardens.

Within a decade the first-phase target of 600 houses had, to all intents and purposes, been reached.

The wide roads are tarred. Residents have a recreation club. There are schools to cater for various age groups, as well as a modern technical and commercial high school. A creche has been established to care for small children while mothers go out to work. A clinic and ambulance services take care of health emergencies, while the main hospital is a few kilometres away at the mine itself.

While a number of sites at Wedela have been set aside for churches, none has yet been built. For the moment, all denominations use the school and community halls for religious services and the various church activities.

The government's scrapping of its policy of influx control in 1986 made it possible to extend the concept of bringing workers' families to the mine. Earlier the law had allowed not more than three per cent of a mine's black workers to be housed at the mine with their families. A poll conducted by Anglo American indicated that of the 58 000 married workers on its mines, 24 000 preferred to bring their families with them, choosing to live in houses rather than alone in hostels for single men.

In October the following year, the Corporation announced that its black mine workers would be given the option of living with their families in their own houses near the gold mines where they are employed. For those who can afford them, thousands of homes for them will be financed by building societies in the usual way - and rentals have been calculated so as to put such houses within the financial reach of the majority of black employees.

In the middle of 1987, Peter Gush, Chairman of Anglo's Gold and Uranium Division and of Western Deeps, announced that "recently legislated reforms have enabled us to develop a new housing policy, our ultimate objective being to give our employees the opportunity of owning their own homes."

In publicising Anglo's housing scheme, a government publication later said that "Anglo American Corporation is working on a major project to give married employees the option of living with their families near the mines and thus reducing the number of workers living in hostels. Elandsrand, in a joint project with Western Deep Levels, has started construction of additional housing units as well as social and other amenities at Wedela in the Western Transvaal. The houses are for lease or for sale to black employees in terms of a subsidised home-ownership plan."

It marked the advent of the era which HFO had envisaged and pleaded for as long ago as 1952 in the face of opposition from the authorities.

Single workers live in four hostels situated at each of the shafts and an additional central hostel for senior black employees. In 1986 some 21 194 single workers were accommodated in this fashion. The employees at Number 1 Shaft, consulted in the design of their new hostels, produced some constructive proposals which, as Mr Gavin Relly has pointed out, have resulted in improved accommodation and facilities.

Max Velani, originally from the Transkei, is proud of the club he manages at Wedela, with its two bars, snooker tables, TV room, recreation halls, tennis courts and Olympic-sized swimming pool. An old hand at Western Deeps, he has worked for 14 years in a variety of responsible positions: at the mine's telephone exchange, in the timekeeper's office, in the mine's post office and, for eight years, in the Western Deeps recreation club. Subscription for membership of the Wedela Club is R1 a month, happily paid by its 285 members.

On the north-facing slope of the Westwitsrand between Numbers 2 and 3 Shafts, on the one hand, and Number 1 Shaft, on the other, where staff houses nestle in gardens among the trees, there is also the Village Club, recently built. Subscriptions are very reasonable and the club is open to all employees on the mine, offering a wide variety of sports and cultural facilities, as well as a splendid restaurant. Managed by Collin Latulipe, it has been described as "the finest club in the (mining) industry."

Funding, amounting to more than R8-million, for the Wedela Technical High School was provided by the Anglo American and De Beers Chairman's Fund. The pupils are intensely proud of their well-equipped school, justifiably pointing out that the computer equipment is more modern by far than that at other schools in the area.

The school was the first of its kind in South Africa, providing blacks with technical training. Opened in 1984, it was the brain-child of

Village Club: recreation Club on Westwitsrand.

Gerald Langton and Rawdon Lawrie, Chairman and Managing Director respectively, who saw an acute need to develop technically literate black children, as opposed to the usual matriculants. It is a remarkable success, and its pattern has been followed by a number of similar institutions throughout South Africa.

Culturally the pupils are proving themselves too. The Wedela school choirs have achieved success in a variety of local and national song competitions, winning trophies and prestigious titles. With pride they talk of appearances on TV in South Africa, and tell of a recent singing tour to Israel which has led to a promise of a visit to Germany.

The choirs are the brainchild of Bax Modise, who is keen to promote both indigenous and western music among the pupils and citizens of Wedela. His wife shares his ambitions and interests and assists enthusiastically in this work.

The workers' township of Wedela has no equal in South Africa — and there is almost certainly none as good on the whole continent. In the Anglo tradition Western Deeps has always been progressive in its approach to housing. Arthur Wilson, Anglo's Public Relations Officer in the fifties, had considerable influence on the standards of layout and design of the first living quarters. The quality and style represented "a tremendous leap forward", as John Shilling recalls, though the concept of the married quarters of those days has now been overtaken by more modern complexes.

A well-equipped hospital at Western Deeps takes care of the health of some 30 000 workers, black and white, and their families. A large modern building set among trees and colourful gardens high on the sunny northern slope of a grassy ridge, it overlooks the Number 2 Shaft complex and, beyond the mine dumps, to the wide expanse of the rippling highveld beyond.

But it did not begin that way. At first, when the mine was younger and smaller, the hospital consisted of only two rooms, part of a building near Number 2 Shaft. The present hospital was planned by Dr Jack van Blommenstein, Medical Consultant for the Anglo American group.

Dr Leontsiris was its first superintendent and Sister Dixie Rouillard its first Matron. Her staff consisted of a handful of black trainee nurses. There were two black doctors.

In the ten years he has supervised the work of this hospital, Dr Pat Gillies has seen many changes and developments, both of people and

Wedela Technical High School.

trends. While years ago the primary function of a mine hospital was the treatment of pneumoconiosis, the dust-induced disease, that is no longer so. Mining engineering has improved conditions underground and the disease has become something of the past. Today the hospital deals with a large number of trauma cases, handling an average of about 200 cases a month, most of which require orthopaedic attention or treatment for shock rather than treatment for surgery. In addition, the hospital handles a number and variety of medical cases.

Between 1973 and 1986 the total labour force at Western Deeps rose from some 16 000 employees to roughly 28 000. Admissions to the hospital — not unnaturally — rose accordingly, though the percentage fluctuation was slight.

Not long ago all admissions were male and were mostly younger, inexperienced workers. But now, with families living on the mine property, the hospital has to deal also with women and children. In a sense it is thus ceasing to become a specialised mine hospital and, dealing with the whole spectrum of health care — including gynaecology and obstetrics — is becoming more and more like the usual general provincial hospital. Increasingly, too, in line with contemporary trends, specialists are taking over from the general practitioner. More and more are specialists becoming resident doctors, rather than itinerant visiting consultants.

Before coming to South Africa Dr Gillies was the Regimental Medical Officer in the Scots Guards. His bearing testifies to a military background. He sees his functions at Western Deeps essentially in the round. Routines involve not only the daily treatment of sickness, disease and surgery, but also in encouraging mass education to prevent and eradicate threats to good health.

Regular classes in family planning, general hygiene and health education are a part of the hospital's routine. The large concentrations of workers and the fact that classes can be arranged to fit convenient times make the mine a useful centre for such education.

Pat Gillies shares the prevailing idea that, with the soaring birth-rate assisted by the elimination of many of the traditional diseases which once held the Third World's population increases in check, the key to Africa's future lies in controlling population growth and overcoming the rural black's frequent resistance to family planning. Diminution in the birth-rate seems generally to be brought about by improved socio-economic conditions. With the amelioration of living conditions, with education and urbanisation and, most of all, with increased earning power, the birth-rate in South Africa will probably begin gently to decline over the next five years or so, as social aspirations develop among blacks. If this proves correct, Western Deeps will have contributed substantially to the future welfare of its black employees.

In his years at Western Deeps, Dr Gillies has seen how swiftly concepts of disease and sickness have changed in Africa. Some resistance to the white man's systems of medicine continues, nevertheless, and tribal beliefs still prevail.

For many blacks illness is still frequently believed to be caused not by something, but rather by someone: it is not what made them ill, but who. Ancient beliefs die hard. Modern medicines still cannot exorcise hostile and evil spirits wished upon men and women by enemies and rivals. The most up-to-date pills and prophylactics remain strangely powerless against the forces of such superstition.

Black patients often prefer to be treated by their tribal witchdoctors than by the mine's physicians. In such cases, those at Western Deeps are encouraged to consult their traditional medicine-men and are given time off to do so. Workers generally return, however, with their psychoses cured but replaced by some dire physical illness caused by the potions of the tribal medicine-man.

The psychological illnesses of the white man seem to be taking hold among blacks, especially among those who have an advanced modern education. Psychiatrists, both white and black, seem powerless to handle such cases, probably because both are trained in the white man's traditional systems and are unable to penetrate the African psyche and ethos. These developments are typical of what is happening wherever the cultures of different worlds come face to face, however. Manifestations of the consequences of this cultural confrontation are part and parcel of the hospital's daily routine.

Inevitably, the new and modern disease of AIDS, rampant in parts of Africa, is watched with apprehension and the uttermost care. Alien workers especially undergo the strictest tests.

The biggest study yet to be carried out on this killer disease is being done by the Chamber of Mines. Begun in 1986, tests have been carried out on every new recruit and on every worker returning from leave. By the end of 1987 more than 300 000 blood samples had been taken. These revealed that, so far, the incidence of AIDS is not high among South African workers. On Western Deeps only three cases of AIDS have been identified – out of a work-force of 35 000. The surrounding area, with several gold mines, has a high density of black population, generally some 150 000 workers at any given time. To date there have been only three deaths from AIDS. Without a known cure and, so far, without an effective antidote vaccine, however, the mines are watching the situation with care and concern.

All possible is done to ease the strains of mining at Western Deeps. Workers have four months' leave a year. Conditions underground are constantly being improved. For example, reduction of dust in the air underground reduces the incidence of pulmonary diseases. Heat is being conquered and controlled, noise is coped with. The health of the worker is of paramount concern. The hospital at Western Deeps is an important element in the whole operation, caring for the present and looking to the future. Dr Gillies says he is "trying to boost mining medicine into a new era."

The maintenance of health and fitness is encouraged, and available recreation facilities cover a broad spectrum. For blacks soccer is the prime sport, its popularity demonstrated by the huge crowds at matches. Athletics are growing in popularity. There are sports for all seasons.

The Westwitsrand has a nine-hole "chip-and-putt" golf course laid out by Andrew Tennent when he was General Manager, together with Felix Parry, the Surveyor, and Dick McCracken, a former South African Amateur champion. Wryly Andrew Tennent recalls the rigorous steps taken to ensure that only indigenous trees were planted there.

The Boskop Yacht Club owes its establishment to two enthusiasts at Western Deeps. It was founded, in 1959, by Dick Lambert, the mine's first Assistant Manager, and Leal Wright of the Planning Department, both keen yachtsmen. From its inception, Dick Lambert says, the Club was popular "with members, friends and families as a place to relax during the weekends from the rigours of bringing the mine to production." In 1979 the *Carletonville Herald* reported the Club's twentieth anniversary. It is still going strong.

While sport makes itself prominent among the staff, there are numbers of clubs catering for a variety of other activities. Western Deep Levels also has its own Scout and Cub packs, for example, with premises of their own. An Equestrian Club, with stables, caters for those who wish to ride, while today's trendy youngsters have a specially laid out course to show their skills on BMX bikes. From among its staff Western Deeps can muster enough sky-divers and parachutists — women as well as men — to form a company team. At the mine's recreation centre there are tennis courts, a swimming pool, bowling greens, a gymnasium as well as facilities for a wide range of indoor sports.

All ages are catered for, with full-time sports administrators helping not only with the organisation of recreation, teams and matches, but also providing coaching and training in various sports. Teams from the mine compete in various Western Transvaal sports leagues, and through the years this has resulted in numbers of its sportsmen and women gaining provincial colours.

Recreation is not limited to physical sport, however. The mine's well-stocked library is greatly patronised. An Adult Education Centre provides facilities for further studies and intellectual recreation. *Western Deeps News* flourishes and has grown into a bilingual, glossy magazine-newspaper produced in five colours.

Besides the creation of handsome sporting facilities at the mine itself, Western Deeps has given generously to the towns of Carletonville and Fochville to encourage sports in the communities where many of its workers live. In 1960 funds were provided for the building of a sports stadium in Carletonville and, later, provided Fochville with a swimming pool which the citizens named after Andrew Tennent.

Personnel at Western Deeps who have worked at other mines, in South Africa as well as overseas, are unanimous that Western Deeps is in every way the most attractive mine property they have ever lived on.

While the local topography certainly lends itself to attractive landscaping, the mine owes its present beauty to the genius of Joane Pim, a landscape gardener whose talents brought her fame far beyond the boundaries of South Africa alone. The title of a book she wrote in 1971, *Beauty is necessary*, sums up her philosophy. She points out that it is not enough that a man "should receive a living wage and have social security". He requires further motivation. Among his needs she lists the requirement of "a harmonious setting to the place of work".

Joane Pim was commissioned to plan and execute the layout of the offices and living quarters at Western Deeps. "A settlement of character and personality has emerged," she wrote, "in the vicinity of two existing shafts and administrative buildings." A settlement of character and personality it is indeed!

She tells how "engineers, planners, architects and the landscape architect exploited the potential for beauty presented, taking advantage of local topography, vegetation, and inherent assets of soil, available rock and paving stone, and shelter from the cold south winds."

The result is there for all to see, for all to admire: the north-facing slope of the Westwitsrand covered with trees that give privacy to houses tucked among them, gardens made bright with shrubs and flowers.

Frank Jantzon echoes the admiration of many when he says, "I have been on mines in many parts of the world. Not only in South Africa, but also in Canada, the United States and Germany. But nowhere — nowhere — is there a living place so filled with character and harmony as this mine village at Western Deeps. It is superb: it is truly beautiful."

Commissioned by Anglo American, Miss Pim left her mark on several South African mines. Perhaps the town of Welkom in the Free State is her most outstanding monument. An echo of the tribute to Christopher Wren in London's St Paul's cathedral has been recorded in a memorial to Joane Pim at Western Deeps. A bronze plaque set among flowers reads: "If you seek her memorial look around you."

She was a determined lady — "of steely will", it is said — who brooked no interference with her plans. The story is told how she insisted that the positions of larger trees had to be precisely surveyed in relation to the houses. Woe betide anyone who moved a plant from its precisely designated spot — even Dick Hartley, the hard-working, much harassed gardener at Western Deeps!

When she died in December 1974, the *Rand Daily Mail* said that at Western Deeps "she turned industrial ugliness into something green and congenial." Her death was a loss, felt not only by Anglo American but also throughout South Africa. *The Star* in Johannesburg praised her "life devoted to shaping the earth for man's use and enjoyment." At Western Deeps her work was carried on by David Dutton, later to become Anglo's chief horticulturist.

The South Mine was not in existence when Joane Pim was busily beautifying the property on the northern side of the Westwitsrand. Number 1 Shaft was established in the decade after her death. But the tradition of beauty continues there too. Morgan Zeederberg, the groundsman in charge of that area, works where, a century before, his grandfather's coaches thundered up and down the valley and where there was a staging post with fresh horses to replace the sweating, foaming teams that had brought the mail coach along the previous stages. The stables, now forlorn and dilapidated, still stand in the valley among the gnarled, dying oaks, decaying souvenirs of a former age.

A sculptor in his younger days, Morgan Zeederberg spent many years with a four-pound hammer and a chisel chipping away at huge blocks of granite to shape the 18-feet-high statues of Andries Pretorius and The Unknown Voortrekker which adorn opposite corners of the Voortrekker Monument in Pretoria.

As he walks about the property, planting clusters of trees here, constructing wind-breaks there, does he hear on the wind the echo of the post-horn, the thud of approaching hooves, the jingle of polished harness and the creak of his grandfather's carriages that plied this route a century ago?

Modern mines are self-conscious about conservation of the environment. All possible is done to limit dust from the dumps and to prevent complaints about pollution of the surrounding landscape. They aim to demonstrate that mining and industry can go hand in hand with conservation. The two need not be in competition. Studies are constantly being conducted into improving the means of re-grassing the waste dumps and efforts are being made to ensure that damaged veld recovers.

Low-lying swampy ground near the Number 3 Hostel complex has been turned to advantage with the creation of a small bird sanctuary, with hides to enable bird-watchers to see the increasing varieties of bird life adopting these ponds and miniature wetlands as their habitat. Joane Pim would have approved.

In conjunction with Elandsrand, Western Deeps has taken the concept of conservation a stage further by developing a small game park in the southern section of the property. A pleasant area, it is secluded from the mine itself, with grassy hills enclosing a narrow bushy gorge and a gentle stream that feeds from the head of the valley. Birds and small animals are encouraged to make use of the protection which the enclosed area of 650 hectares provides. Several species of wild herbivorous animals have been introduced there, while many species of native birds are beginning to nest in this sanctuary too.

A picnic area, built of local stone, blends gently into a fold of the hillside, shaded by indigenous trees. A bronze plaque set into the rock tells that this "lapa" was inaugurated in May 1986 by the New Zealand Cavaliers during their rugby tour of South Africa. It is used almost daily for outdoor parties and will become even more popular when the walking trails in the nature reserve become better known.

Given the variegated social make-up of South Africa's population, the racial differences and the gaps — now closing — in education, it is not surprising that industrial relations in the country are complex and fraught with difficulties. Harmonious management/worker relations are essential in the social, economic and political development of the country and, if recent developments and events are any indication, the importance of understanding and harmony will increase with time.

Inevitably the gold-mining industry, as the focal point of the nation's economy, is sensitive to breaches in such harmony. So important is the smooth running of the mines to the economic life of the country, that it is only to be expected that it will be used for political ends — from both within the country and without — by those eager to hamper national development.

Labour relations are coming increasingly under the spotlight in South Africa. In 1977 the Minister of Labour and Mines appointed a Commission of Enquiry to examine the situation and to recommend legislative changes to improve the handling of labour matters. The chairman was Professor Nic Wiehahn and its report was published in 1980.

In the company's own annual report for the following year, the Chairman of Western Deeps, Gerald Langton, told shareholders that the mining industry "... has embarked on discussions with the trade unions and officials' associations which are intended to provide white employees with a greater feeling of job security without which the desired improvement in the utilisation of all employees cannot be achieved.

"Meanwhile, trade unionism amongst black workers, which has grown rapidly in the South African commercial and industrial sector, is showing signs of emerging in some areas of the mining industry, though no formal approaches for recognition have been received by any of the companies. . . ."

Referring specifically to Western Deeps, Langton went on: "... For many years we have been engaged in improving communications and relationships with our employees and creating better living and working conditions. A system of industrial relations audits which determines the extent to which the mine has achieved the targets it sets for itself has been introduced . . .

"Slow progress is being made in the sensitive interface areas of our industrial relations because of an inherent conservatism and the fact that there are such large numbers of workers on the gold mines who come from diverse parts of southern Africa with different languages and cultural backgrounds, the majority of whom, generally, have not been afforded the opportunities of more than primary education. This makes it apparent that a smooth process of change and development faces formidable difficulties. A further constraint is that of finance, particularly as the costs of housing, amenities and infrastructure have escalated so alarmingly in recent years that it is not always possible to proceed as fast as we would wish.

"Nevertheless, important changes have taken place on this mine and in the mining industry as a whole over the last few years and a good foundation is being laid for the future."

Western Deeps was not without its share of labour problems, some, however, unconnected with conditions on the mine and unrelated to dissatisfaction with aspects of the administration. In 1974, for example, as a result of violent faction fighting and political agitation, eleven men died and 101 were injured. The Chairman noted that "... the causes of clashes between different ethnic groups remain obscure, and hence it is extremely difficult to arrive at a solution ..."

Differences of opinion between employees and employers have often been due to misunderstanding — or lack of understanding — of the motives of management. Some years back, Anglo's mines embarked upon a policy of closer consultation between worker and management, and went to considerable lengths to explain intentions before they were implemented. In 1981 the Corporation enunciated the principles to which it adhered in respect of industrial relations and employment practices.

Together with a summary of progress made in their application at Western Deeps, details of those principles and practices appeared in Western Deeps' annual report for 1981:

> "The mining companies administered by the Anglo American Corporation of South Africa seek constructive relations with unions representing employees both at industry and mine level. It is considered desirable that collective bargaining is co-ordinated to the extent that matters affecting the industry as a whole are dealt with at industry level, through the Chamber of Mines, whilst mine-level issues are dealt with at the mine.
>
> Discussions are taking place with the Council of Mining Unions and the officials' associations in order to improve the industrial relations structure within the industry, provide adequate security both for individual employees and the organisations which represent them, and to debate the principles underlying the existing closed shop agreement.
>
> At present no formal approaches have been received by unions seeking to represent the black employees of the various mining companies administered by the Corporation.
>
> Company-level structures exist in order to promote consultation and the solving of worker problems. These consultative committees represent the majority of workers to the extent that a diverse workforce of this nature involves itself in the election of committee members. Formalised systems operate to ensure that relevant information reaches all employees. A system of Briefing Groups is used for this purpose. While this has worked well, there have been

breakdowns at some levels. In order that discipline should be fairly and consistently administered, and that employees have the opportunity to raise and resolve problems, discipline and grievance procedures have been established for some years.

Little change has taken place in recent years in the manning structure of the industry and progress in this area will be dependent on the evolution of a more satisfactory industrial relations attitude and the development of a sense of greater security amongst white employees.

Considerable finance and effort is devoted to the training of employees, both to allow them to become more productive and also to help allay fears held by many concerning their future security in the mining industry. The Employee Development Programme, which is being introduced, is aimed specifically at the personal development of employees.

This mine, in common with other mines administered by the Corporation, has adopted the target of achieving a minimum cash wage which will allow the migrant miner to provide for all his, and his family's, basic requirements. Important progress has been made towards this target during the year under review.

Whilst moves continued towards the provision of common or equivalent benefits for all employees, the reactions of some blacks in secondary industry to the proposed preservation of pensions, as well as the reaction of some blacks in the mining industry to the introduction of the death benefit insurance scheme, emphasise that any change to the present range of benefits offered to employees needs to be introduced with caution and then only after thorough consultation with all those affected.

Ideally, all employees should have the opportunity of living at or near their place of work with their families if they so choose. Where Government policy or financial constraints make this impossible, and the company is responsible for accommodating employees in single quarters, every effort is being made to ensure an acceptable quality of life. Although many of the men living in our hostels are single, the larger proportion is married, and to lessen the undesirable consequences of the migratory labour system, major efforts are being made to provide higher standards of accommodation and recreation facilities, flats to enable wives of employees to visit them and to encourage employees to make more regular weekend trips to their homes, where this is practical. During the year, R8,6-million was spent on improving the amenities for all employees. The bulk of this expenditure was incurred in improving hostel accommodation, dining-halls and kitchens.

A modern village, Wedela, has been established where senior black employees of both Elandsrand and Western Deep Levels are housed. To date, 364 houses and a primary school have been built at a total cost in excess of R7-million. A further 70 houses will be built during 1982.

Hospital facilities for black employees have been improved with the addition of such facilities as an intensive care unit, an obstetrics unit and laboratory and pharmacy facilities.

Since the inception of a home ownership scheme, 128 employees have made use of the scheme to purchase their own homes, of which 52 were formerly company-owned houses.

During 1981, 26 houses for whites were built, whilst a further 14, together with 96 flats, were purchased. It is intended to build an additional 25 houses and 52 units for single officials during 1982."

The intention, there expressed, to achieve a minimum cash wage sufficient to provide for the basic needs of a miner and his family, has meant, of course, regular upward revision of his pay in order to keep pace with inflation.

At Western Deeps this has been done. Workers' wages have kept ahead of inflation.

The building of houses for black and white workers did not cease with that report in 1982. The building programme has continued — and still continues — so that by the end of 1987 the company's Chairman was able to report that there were 590 houses in Wedela — all occupied, many owned by the employees themselves.

CHAPTER ELEVEN

The Seventies and beyond

For Western Deeps, now firmly established, the seventies began quietly, but with an optimism fuelled by the buoyancy of the rising gold price.

As the 1960s drew to a close, John Shilling had taken over from Hermann Koch as Chairman. No stranger to Western Deeps, he had been the first Managing Director and had seen the launching of the "greatest enterprise in South Africa". He had been a member of the Board from its inception.

Shilling's first report to shareholders, covering that final year of the sixties, confidently outlined the positive progress of that year, leaving the statistics to emphasise the details of achievement. His report for the following year was similarly factual — and so was the next. And the next. There seemed little to disturb the even tenor of progress made manifest from figures — impressive figures — of metres stoped and exploited, of shafts sunk further, of gold produced — in 1971 at higher prices — and of profits which, despite continuing inflation, were now bringing the "tax honeymoon" to an end.

For several years shareholders had been advised that "... the total redeemable capital expenditure incurred by the company in excess of the aggregate profits earned is treated as an assessed loss for taxation purposes, and no taxation or State's share of profits will be payable by the company until this assessed loss has been absorbed by profits." The statement had become a formula.

In 1972, however, ten years after the production of the first gold, the Chairman told the AGM that "... it is likely that the company will be liable for some taxation in the current year ..." At the close of the previous year the assessed loss for taxation purposes was provisionally calculated as being R14 940 000.

Miners coming off underground shift at Western Deep Levels.

His supposition was not wrong. For the year 1972, Western Deep Levels paid R9 595 000 to the State in taxation and, from then on, to coin a phrase, the government never looked back! By 1986 the company had paid no less than R1 618 904 000 in tax to the State. It would be only fair to add that at that stage shareholders had been paid a total of R1 328 240 000 by way of dividends.

Released in 1969 from the constraints of an internationally fixed price, gold began to reflect a steady increase in market values, so that soon — early in 1972 — gold was selling at $48 an ounce. Not only did this substantially increase profits for the producers, but it also had the effect of forcing up the official gold price, thus strengthening the position of bullion in the international currency system. The uncertainty and the concern of the world's money markets, shaken by the prevailing oil crisis, became apparent. World prices of gold fluctuated considerably in the following months, reaching almost $95 an ounce a year later.

The abandonment in December 1973 of the two-tier system of marketing gold and the consequent termination of South Africa's gold agreement with the IMF led to further substantial increases in the market price and in less than two months the price had surged to a record $178. During 1974 the market price increased by 63 per cent.

It was as well for the industry, since rising costs in every section of production were becoming increasingly difficult to arrest and control.

At Western Deeps work was now concentrated on developing the Carbon Leader below Level 100. New longwall stopes were established

Miners transferring from one shaft system to another underground.

and exploratory drives continued at that level. In 1971 at Number 2 Shaft a pilot winze was sunk to 112 Level while the main hoisting incline had reached 113 Level. At Number 3 Shaft a pilot winze had reached 114 Level — and a little further — and there the main hoisting shaft was sunk to 115 Level. At both shafts work on the winders for the tertiary vertical shaft systems was complete and the further shaft sinking began.

Amid great jubilation, April 1973 saw the addition of yet another record to Western Deeps' catalogue of achievements when, as the Johannesburg *Star* reported, "a vertical mine shaft being sunk by Shaft Sinkers (Pty) at Western Deep Levels mine near Carletonville has already exceeded the depth of the deepest mine in the world."

On its way to the ultimate depth of 3 618 metres, the Number 3 tertiary production shaft reached a depth of 3 472 metres below surface, beating the previous world's record depth of 3 428 metres reached at the ERPM mine near Boksburg.

But if underground development at Western Deeps was continuing satisfactorily, similar steady progress was not reflected in all aspects of operations. The next few years were to see the mine struck by a number of problems, by accidents and setbacks of varying severity and consequence, over some of which the administration itself had little or no control, nor for which was it responsible.

On 13 October 1972 a freak explosion on 90 Level of Number 3 Shaft killed 12 miners when explosives were being carted into the working area. The ensuing fire made it necessary to seal off the area for several weeks until the fire had burnt itself out. Lack of evidence and the absence of any witnesses made it impossible to pinpoint the cause.

Almost a year later, in 1973, there was a severe setback of another sort. It came dramatically on the evening of 11 September, when workers living in the mine compound at Number 2 Shaft began rioting. Black machine operators who, dissatisfied with recent pay adjustments, had gone on strike the previous week, tried to prevent the night shift from going underground. Aggressively they instructed the workers to return to their hostels, intimidating those who refused. The threats were backed by violence, which led to malicious damage. Windows were smashed. As the severity of the violence escalated, the mob broke into stores, looting food and, from the compound canteen, also liquor. Quickly the word got about that a strike had been called and the situation, deteriorating fast, was soon completely out of hand. Demonstrators threatened the lives of the compound manager and the mine security officials. At that stage, on the instruction of the Mine Manager, Algie von Holdt, the police were called in. On arrival they were met by a barrage of stones from a savage mob now brandishing a variety of dangerous weapons. Both teargas and, later, a baton charge failed to subdue or disperse the rioters. In the prevailing darkness, the mob, now grown to several hundreds, surrounded the handful of police who were forced to use firearms.

Eleven black miners died. Another was bludgeoned to death by rioting workers.

International repercussions were swift. The shootings were given prominence in the media around the world and, inevitably in the prevailing international anti-South Africa climate, comparisons were made with the Sharpeville incident thirteen years previously. Prices of gold shares dropped on the Johannesburg Stock Exchange, and overheated selling of South African shares overseas brought prices down sharply, though bullion was traded at higher levels. In Australia dockworkers refused to handle South African cargo. In London the Internationalists organised a protest against South Africa. Demonstrators assembled at Hyde Park Corner for an organised protest march on the South African Embassy. This turned out to be a damp squib. Its path crossed that of a larger and separate demonstration — a crowd which happened then to be marching to the Chilean Embassy to protest against the military Junta that had toppled President Allende. The ensuing confusion turned the anti-South African march into something of a farce.

Greatly outnumbered by their anti-Chilean counterparts, those protesting against events at Western Deeps simply tagged on behind the larger column and, thus deflected, marched willy-nilly to the Chilean Embassy — "as a demonstration of solidarity", not, they later said defensively, for any reasons of lack of support in the move against South Africa.

The international diary schedules September as the month when the General Assembly of the United Nations regularly convenes in New York each year. Seasoned UN campaigners and political organisers everywhere know that each September thus brings with it fresh opportunities to air grievances which might be seen to have an international flavour. With all the world represented in that forum, the broad spectrum of political interests ensures that there are always some who are readily available to raise contentious issues in the General Assembly. The UN thus provides a useful platform for militants of many persuasions seeking the spotlight.

While there is no suggestion that the "Carletonville shootings" were inspired from abroad, they nevertheless provided the UN Apartheid Committee with yet another opportunity to drag South Africa into the international dock. It did so without hesitation and the ensuing debate was seen to be too urgent to await any assessment of facts which would emerge in the course of the official enquiry in the magistrate's court at Oberholzer.

On the afternoon following the riot, Western Deeps issued a statement that operations at the mine were back to normal. This did much to defuse the situation, and the gentle upward movement of gold shares reflected the Stock Exchange's return to normal.

On the BBC's current affairs programme, "Panorama", that week Mr Harry Oppenheimer, Chairman of Anglo American, expressed deep regret at the loss of life. Discussing the various causes of the disturbance, he stressed how rioters had incited others to join in a strike about pay and, in doing so, had created a situation which endangered life and property.

This was confirmed by several workers. In the enquiry which followed, they told the presiding magistrate how they had been attacked, beaten up and stabbed by militant rioters because they had refused to take part in the fracas.

The magistrate found that neither the police nor the mine authorities were to blame for the killings; and he laid the responsibility "foursquare on the shoulders of the rioters".

Western Deeps is regarded as a model mine. Discussing that night of ugly rioting, the Chairman said, "... The fact that this all happened on a modern mine whose black workers are amongst the best paid in the gold mining industry, and whose conditions and amenities are generally of the highest standard, added to the complexity of the situation ..."

That complexity led to much soul-searching and, in turn, to in-depth examinations of labour relations in the mining industry.

This was the first of a number of serious labour conflicts which have plagued South Africa's gold mines in recent years. In a scholarly study of the problem, Dr Kent McNamara, an Industrial Relations consultant then attached to the Chamber of Mines, has probed deeply into the reasons for the disturbances.

He has said that while the riot at Western Deeps was not wholly representative of conflicts which took place subsequently, "... it came to be regarded by industry, management and the public alike as the symbolic end of an older era of labour relations and the beginning of a new chapter in the mining industry's history. In popular terms, the Carletonville shootings marked the appearance of a new tide of black anger and frustration in a reputedly docile and submissive industrial labour force. . . ."

Most conflicts on the mines seem to stem from questions of pay. The September riot at Western Deeps proved to be no exception. In this instance, a complete overhaul of wage structures, evaluating responsibilities of various categories of work in 1972, had led to new differentials in the wages paid to certain categories. In this reappraisal of duties, some increases were proportionately greater than others, resulting in a narrowing of the wage gap between certain classes of workers. This was resented by workers in those categories which appeared to have been left with diminished status.

Machine operators comprised one such category. It was they who had refused to work and who sparked off the bloody riots at Number 2 hostel in September 1973 while attempting to involve other workers in their grievances.

The workers' lack of understanding of the procedures involved was unfortunate indeed. From the start, in re-assessing work responsibilities, management had gone to considerable lengths to inform workers of the reasons for the new system. Notwithstanding the extent to which management had gone to ensure that the workers understood the rationale of the new system, there were clearly some who failed to appreciate what was being done.

In his annual report, outlining the steps taken during the difficult period of change, the Chairman told shareholders, ". . . Considerable attention had been given to establishing more effective means of communication to ensure that management was fully aware of the views of its workers and they, in turn, of the intentions of management. An elaborate system of communication committees had been established at the mine but it was fairly new and largely untried at the time of the disturbances . . ." Clearly, however, these efforts were only partly successful. They were hampered by the diversity of the backgrounds of the black workers and the fact that circumstances make the implementation of new procedures a lengthy business. Closer communication was later established, however, when a consultative committee structure was developed whereby elected blacks could "communicate more effectively to management the collective ideas and views of black employees."

In August 1971 Western Deeps was honoured by a visit by Dr Hastings Banda, President of Malawi. While on a State visit to South Africa, he was flown to the mine one afternoon in a military helicopter, accompanied by a number of craft carrying other VIPs, among whom

Visit of President Banda of Malawi to Western Deep Levels; August 1971. Left to right: *Algie von Holdt, Manager of WDL; John Shilling, Chairman; President Banda; Dr Carel de Wet, Minister of Mines.*

were Dr Carel de Wet, Minister of Mines, and John Shilling, Chairman of Western Deeps.

Tables laden with sandwiches, cakes and the paraphernalia of a welcoming tea party were set at the edge of the sportsfield where the cavalcade of helicopters was to land. Unfortunately one of the descending craft approached too close to the reception area and the mini-gale of its rotors swept over the table. Cups and saucers, plates of cookies and sandwiches crashed to the grassy earth, ruining the party prepared for the President who went tea-less to address the huge crowd of mine-workers — Malawians as well as others — excitedly awaiting him in the mine-dance arena.

It was by no means President Banda's first visit to a South African mine. As a young man he — like thousands of other blacks — had left home and walked more than a thousand miles down the continent to seek employment and a living which the rest of Africa then could not provide. That job on a South African gold mine, almost fifty years before, had been the first modest rung on his climb to world fame.

Later generations of Malawians also flocked to the mines of the Transvaal, coming in their thousands. But no longer did they face the long journey on foot. Formally recruited by the labour wing of the Chamber of Mines, thousands and thousands of foreign black workers have been brought safely to "Egoli" by air. In April 1974, however, an aircraft belonging to Wenela, the Chamber's recruiting agency, crashed shortly after take-off in Botswana, carrying time-expired miners back to Malawi. The crew and seventy-four miners lost their lives. It was the first fatal accident in the history of Wenela's busy air service.

At once President Banda forbade further recruitment of Malawians for the gold mines and continued with the ban until 1977. This embargo brought home to the mines the extent to which the industry was dependent on foreign labour and prudently steps were taken to safeguard against any further such actions. As a result, when the ban on their recruitment was lifted, Malawians found very few vacancies on the gold mines, and the Malawian work-force was reduced to roughly one-eighth of its former size.

Western Deeps was no exception. It was a matter of regret that sufficient vacancies could not be found for the Malawians who wished to work there. They have a fine reputation for reliability and a worthy record of industriousness. The loss of revenue to Malawi was considerable. In retrospect, Dennis Etheredge, Chairman at that time, regarded the ban as unfortunate both for this country's mines as well as for Malawi, with whom South Africa had — and has — close diplomatic ties. The shortage of labour throughout South Africa's gold-mining industry during the following year influenced production and efforts were made to resume recruitment in Malawi.

Negotiations were set back, however, when a Malawi miner was stabbed to death at a film show when an inter-tribal fight broke out at Western Deeps. Several hundred of his compatriots demanded to be sent home, while Malawian workers on other mines also sought repatriation.

Traditionally Lesotho has contributed substantially to the labour force on South Africa's gold mines and the earnings of these workers has long been a major source of revenue to that little mountain state. Early in 1975 the Lesotho Government introduced a system of compulsory deferred pay and the resentment among Basotho workers led to work stoppages at Western Deeps. A delegation of twenty representatives from the mine went at once to Maseru where negotiations brought about changes making the scheme more acceptable to the Basothos. Not long afterwards, however, a number asked to be repatriated after inter-tribal fighting broke out at Number 3 Shaft.

The year 1974 was bedevilled by a variety of labour problems throughout the industry, many caused by tribal friction wholly unconnected with any mine or mining.

The adverse impact of labour disruptions upon productivity leads each time to further consideration of the advisability of increasing mechanisation in an attempt to minimise — or to eliminate — costly stoppages. The argument in favour of mechanisation is further strengthened by the sharply rising cost of labour.

It is a debate with many facets, a problem with infinite thorns, the economic argument weakened by the countervailing forces of social responsibility in a country in which the mining industry is the nation's biggest employer. Many millions rely on the mines for their existence. But such responsibility must be reciprocated. It cannot be one-sided.

The labour disturbances at Western Deeps, which attracted international publicity in the mid-seventies, led indirectly to a further rumpus a decade later. It was an unusual and bizarre affair in which inaccurate information fed to the media caused South Africa — and Western Deeps — considerable embarrassment.

In October 1985 Mr Malcolm Fraser, a former Australian Prime Minister, said in the United States that at least one South African gold mine had installed a system to pump disabling gas through mine ventilation shafts into the living quarters of black mine workers.

He declined to name the mine, and the storm which erupted in the world press was accompanied by a storm of speculation in South Africa.

The Chamber of Mines described the allegation as nonsensical and denied knowledge of such a system, going on to point out that mine ventilation shafts formed a functional part of mining operations and in no way were connected with employees' living quarters.

Anglo American issued a similar denial in respect of its own gold mines.

Peter Gush, Chairman of Western Deeps, then issued the following explanation:

> "During serious inter-tribal rioting at Western Deep Levels gold mine in 1974 and 1975, several workers lost their lives and considerable damage was done to property.
>
> The situation was seriously aggravated after rioters looted the mine liquor stores. As a result, Western Deep Levels decided to install tear-smoke systems in the liquor stores, hostel kitchens and the administration offices, which could be activated by management.
>
> The system, which used conventional tear-smoke and was introduced as a precaution to protect lives and property in the event of unrest, has not been used since 1975.
>
> Accordingly, allegations suggesting that a system exists to pump disabling gas through ventilation shafts into black mineworkers' living quarters are unfounded."

Management at Western Deeps remembered that these anti-riot protection devices had indeed once been installed at the mine. So long ago had the system been used, however, that its existence had been forgotten. Gencor and Anglovaal, which had admitted installing similar tear-smoke systems, had not ever put them into operation. At once these safety control measures were dismantled.

Malcolm Fraser had visited South Africa not long before on a one-man UN mission to study the operations of multinational corporations in the country. At the time of his allegations in the US he had been designated as a member of the Eminent Persons Group — "the Three Wise Men" — nominated by the Commonwealth to visit the Republic to negotiate with Pretoria about ending apartheid.

Given his well-known hostility to South Africa, many saw Malcolm Fraser's utterances as deliberately malicious, inferring that what were essentially protective devices at these mines were really intended to be used as aggressive weapons against the workers.

During 1976 two major underground fires further upset the rhythm of production. One broke out on Level 109 in Number 3 Shaft in March and was not extinguished for three months. Not only was work impossible during that time, but when the fire was snuffed out weeks had to be spent on rehabilitating damaged areas and equipment before operations could be re-started. Another fire, discovered on Level 94 in June, led to the deaths of five members of the proto team from the Stilfontein Gold Mining Company before the blaze was brought under control and eventually extinguished.

Every miner is aware of the danger of an underground fire. It is a fearsome thing: always costly to production and always a threat to life. Every precaution is taken to prevent them. Early in its life Western Deeps took preventive measures with the installation of "tell-tale" devices in the return airways to give immediate warning of smoke, carbon dioxide and carbon monoxide. Timber used underground is treated beforehand to make it fireproof and standards of safety precautions against fires have been defined by the Chamber of Mines.

Alton Williams, with long service at Western Deeps, now a senior manager, says it is only a miner who can really appreciate the terror of an underground fire.

But, he adds, "it is the trauma of fires, of rockbursts and underground accidents which brings out the most amazing character in people." And the team spirit in rescue operations — "all for one; one for all, regardless of race, is something which has become a part of the particular charisma of Western Deeps," Williams says.

With pride — pride unquestionably justified — Fred Bayley says: "At Western Deeps we have never had to call for volunteers to man a rescue team. As soon as there is news of an accident, men rush to the shaft to make themselves available to help their comrades in distress. They come without being called."

Today rockbursts continue to be the main hazard facing miners underground. Strenuous efforts are being made, with encouraging progress, to monitor their approach and limit their threat. The menace is being overcome, and evidence indicates that present conditions at Western Deeps are better than those in the past. There are many who have experienced the shock and horror of rockbursts.

In June 1972, for example, a particularly severe burst which killed six miners at Number 2 Shaft, trapped a seventh, Maseru Lekhelebane, a Mosuto "boss-boy", who was pinned down by fallen rock that held his leg in a vice-like grip. He could not move or be moved. All efforts to free him proved in vain and, to save his life, there was no alternative to amputating the leg.

To perform the operation, Dr Brian Parker and an ambulanceman, "Corrie" de Jager, both young and both of Western Deeps, crawled 18 metres through a rough and very restricted opening only 45 centimetres high. Extremely hazardous conditions were made more dangerous by rock falling continuously from the cracking ceiling above them. In this confined space and in intense heat, which, from time to time, threatened to overcome them both, they worked desperately for more than an hour to free the trapped miner. Threatened by another rockfall, they were ordered to withdraw immediately. In the nick of time. Minutes later there was another rockfall. As soon as possible, the doctor and the ambulanceman made their way back again along the confined, hazardous passage, sadly only to find the trapped miner dead.

For this extraordinarily courageous effort, performed in the highest traditions of service, in January 1973 Brian Parker and Corrie de Jager were presented with the Chamber of Mines' Award for Outstanding Bravery.

In each case the presentation consisted of an illuminated citation and a gold watch, appropriately inscribed.

In making the presentation, Jeff Goode, President of the Chamber of Mines, lauded their selfless bravery and went on to say: "... These two gentlemen are not mining men and the conditions under which they laboured must therefore have been all the more frightening and horrifying. Yet they knew their duty — to save life if this were possible and if others acquainted with mining conditions thought the chances worthwhile ... They both showed courage of a high order. For them, with so much to live for, it could not have been an easy decision to go into that dangerous place, to stay there as long as they did, and to carry out an operation under just about the worst conditions that can be imagined. ... Yet this is what they did, without hesitation, and in so doing they have added another example to the many exploits that have built a tradition in mining of going to the help of a fellow worker, no matter what the danger ..."

Brian Parker has since left the world of mining medicine, happy now to be a country doctor in the Cape. Corrie de Jager, however, works still at Western Deeps, now the Senior Medical Station Superintendent at Number 2 Shaft.

On the back of the gold wrist-watch he wears is engraved:

C.L. de Jager
vir dapperheid
op 19 Junie 1972
te Western Deep Levels.

On the watch face there is the inscription "Chamber of Mines Award for Bravery".

Each also received a cheque from the John Montgomery bequest to the Chamber of Mines "for the most outstanding act of heroism whilst engaged in mining operations or activities."

These were not the first acts of bravery at Western Deeps to be recognised: nor the last.

In December 1969 George Jagals, a timberman at Number 3 Shaft, had rescued a black worker, Bassilio Munquambe, who had slipped and fallen about 40 metres down an ore pass. The lack of response to shouts led colleagues to conclude that Bassilio had either been killed or knocked unconscious.

George Jagals, fully aware of the risks involved, volunteered to be lowered on a rope into the ore pass to investigate. He found Bassilio unconscious, with a severe head injury and multiple injuries to the body. Air hoses were lowered to combat the unbearable heat. Jagals tied a rope around the unconscious man. On second thoughts, however, the seriousness of the injuries led him to decide that, rather than hoist the worker, it would be better and safer to drain the ore pass of rock so that Bassilio could be lowered through the chute to the level below.

To give support to the injured man, Jagals then roped Bassilio to himself and, with the unconscious man on top of his own body, he guided their bodies down to the chute position as, slowly, the level of the broken rock was lowered. "In other words," as the President of the Chamber related at the presentation ceremony, Jagals "converted himself into a human stretcher in order to convey the unconscious man to safety without further injuring him in the process."

Bassilio recovered fully from his injuries.

The gold watch for bravery presented to George Jagals was the ninety-ninth such award by the Chamber of Mines over the previous fifty-four years.

At the presentation, the President spoke of the pride and appreciation throughout the mining industry "of the great *espirit de corps* which exists between European and Bantu underground when it comes to the saving of a life . . . Time and time again Europeans have risked their lives to save Bantu; and the Bantu have risked their lives in the same way to save Europeans. . . . ."

The Chamber of Mines Award for Bravery is not given lightly. It is the bestowal of special recognition of a particular deed requiring great courage on the part of a miner who voluntarily puts his own life in jeopardy.

As Peter Bosman, General Manager of the Chamber, said, "... an act of bravery is not something you can train for or work towards, like a promotion. It is not done for personal gain, but as an unselfish act. Concern for a fellow worker in need is the only consideration ..."

Doubtless there are many acts of great courage underground which go unnoticed.

"... Those who feel that bravery in our business is commonplace will find many a sympathetic ear," said Bill Nienaber, Mine Manager at Western Deeps in the 1970s. "Many a hero goes unsung."

Those whose deeds are recognised represent the many unnamed heroes of the mining industry.

At Western Deeps in April 1971, Charles Palm, a foreman fitter, and Paul van den Heever, a shaft foreman, risked their lives in rescuing two black miners who had fallen down an ore pass in Number 2 Shaft.

Charles Palm was lowered by rope into the pass. About 20 metres down, he found one of the miners in a semi-conscious state. While fastening a rope to the victim, he heard heavy breathing and discovered another worker under the rock. Only his feet were protruding. Managing to extricate this unconscious worker, he fastened a rope around him. Palm was then joined in the constricted ore-pass by Paul van den Heever. Each supporting an injured miner and with legs braced against the side of the ore pass, they gradually lowered themselves while the rock was drained away until the box was empty and the victims could be attended to.

Both Charles Palm and Paul van den Heever were presented with the Chamber's prestigous Award for Bravery.

More than a decade and a half was to pass before further awards came the way of Western Deeps, when Rudolph Vorster risked his life to save a colleague, Charles Croucamp, at Number 2 Shaft. In attempting to clear rock jammed in the ore pass above 97 Level, Croucamp was struck by falling rock and, overbalancing, fell down the pass.

Voluntarily and at once Vorster went to the rescue. Lowered by rope, he found Croucamp's body under rock and mud 87 metres down. Unable to move the body and after an hour's struggling in the most dangerous and difficult conditions, he was withdrawn when threatened by heat exhaustion. For his courage Rudolph Vorster was presented with the Chamber of Mines Award for Bravery.

At the same time, Awards for Meritorious Conduct were presented to two others at Western Deeps: Paul Haarhoff and Rudolph Opperman who, with proto breathing apparatus to aid them, braved the atrocious conditions to recover Croucamp's body.

While in no way whatsoever denigrating or under-rating the magnificent courage of those who risk their lives to save others, many feel that the system of individual awards is an invidious one. It is argued that such saving of life — however heroic — is almost always a team effort, and that it is therefore somewhat unfair to reward one without recognising other members of the support group.

There are endless pros and cons to this matter.

In mining — as in every occupation — the hazards and dangers must be seen and accepted as a part of the daily task. This philosophy sets exceptionally high standards for whatever recommendations are made for awards and only the most outstanding are considered.

In addition to those Awards for Bravery and Meritorious Conduct made by the Chamber of Mines, Western Deeps has its own system of recognising acts of outstanding courage, though recommendations are similarly scrutinised with the utmost care and are judged by the highest criteria, and few awards have been made.

Despite the manifold difficulties and problems which bedevilled the seventies and early eighties, Western Deeps continued to make splendid progress, success being measured not only by the gold produced and reflected in the profits it yielded, but also in the considerably improved facilities for the workers at the mine. There was no diminution of effort, and the quest for further expansion never ceased.

In September 1974 the company announced it had joined Western Ultra Deeps Limited in the formation of a new company, the Elandsrand Gold Mining Company, to mine 515 hectares to the south and west of Western Deeps. Western Ultra Deep Levels held the mineral rights over the area, together with Witwatersrand Deep Limited which had a 22,5 per cent participation in the northern portion. In return for the granting of 19,6 per cent participation in the initial equity of Elandsrand, Western Deeps would cede its ground to the new company. By agreement, Elandsrand would mine only the Ventersdorp Contact Reef and Western Deeps would thus retain the right to mine the Carbon Leader throughout the area.

Dennis Etheredge, Chairman of Western Deeps then, has told how exploration of that area by the Corporation's Consulting Engineers had earlier yielded no positive conclusions. "We were not satisfied that we had a mine there," he said.

Gold Fields were mining the adjacent property, Deelkraal. Mr Ian Louw, Chairman, had suggested to Mr Oppenheimer that Deelkraal be linked with Elandsrand and that the two mines operate jointly as one. Just at that time, however, Anglo's exploration drills struck better pay-ore and the borehole results made it clear that Anglo there had a mine of its own. Anglo thus decided to go it alone. A later suggestion that Elandsrand be linked with Western Deeps was similarly rejected, though in the initial stages of its development there was considerable co-operation between the two and both mines fell under the control of a single Regional General Manager. They still do. The first ore from Elandsrand was taken out through the shafts of Western Deeps.

Shaft sinking began at the new mine in January 1976, starting along a road which, it was calculated, would lead to production five years later. Elandsrand in fact reached production well ahead of that estimated schedule, an achievement which was applauded at the time but which led to criticism later when it was found that the problem of inadequate hoisting capacity, limiting output, was exacerbated by the fact that ore values turned out to be less than anticipated. Those values have since improved, however, and the technical snags have been eliminated.

Confidence in the long-term future at Western Deeps was made further manifest by a decision early in 1980 to extend the mine, despite the wide variety of endless nagging problems of the seventies: strikes, riots and other labour troubles which distracted management and plagued development. With the gold price pushed to new heights by political tensions in the Middle East, Western Deeps announced its intention of developing a third shaft. The idea, which had been mooted for some while, was an "exciting new development which profoundly affects the future of our mine," Gerald Langton told shareholders in February.

The proposal envisaged ". . . a twin shaft system which, with all ancillary surface and underground works, including the gold plant, constitutes the first phase of the project. . . ." shareholders were advised. An application for cession of the mineral rights in that area to Western Deep Levels was formally agreed to by Western Ultra Deeps Limited and work on the project was soon under way. This was the genesis of the great Number 1 Shaft which came to production in 1986.

The decision was well timed. Western Deeps had money to spare for, by any standards, 1979 had been a remarkably good year for the industry. Western Deeps had shown a profit of more than R290-million, of which R136-million was paid to the State in tax. The prospects for gold seemed never to have been better. Events in the past year had further entrenched the monetary role of gold in the world's economy.

As 1979 opened, the gold price was a prisoner of circumstances

The beginning of Number 1 Shaft: 1981

created by the US Treasury's policy of increasing the amounts of gold auctioned in support of the dollar. A series of dramatic unforeseen events, however, freed the price from such thraldom.

Five weeks later, following the revolution in Iran, the price of gold soared to a record $254 an ounce in the wake of universal fears of an escalating shortage of oil. By mid-year the realisation of those fears had pushed the price of the metal yet higher and the anchor role of gold in the world's economy was made more apparent by the scramble to invest in bullion.

The market price took a further leap with the mob take-over of the

American Embassy in Tehran in November, and at the close of 1979 it stood at $524, having almost doubled in the course of the year. The climb did not halt there, however.

Only a few weeks into 1980 the sudden Russian invasion of Afghanistan pushed the figure to an incredible $850 an ounce. Though only briefly sustained, this price gave a new impetus to the concept and value of gold.

It looked as if the new Number 1 Shaft was being born with a golden, rather than any traditional silver, spoon in its mouth!

Topography tends to isolate this shaft from the rest of Western Deeps and its shafts to the north of the ridge. On the surface it seems to be a separate mine. Indeed, when its development was first under consideration, there was debate as to whether the shaft should not become a mine on its own, a separate company independent of Western Deeps. "There were many top people," Dennis Etheredge has said, "both in the Company and in the Corporation, who wished to see another mine of the Anglo stable listed on the Stock Exchange." But the advantages of being able to finance the development of the new shaft from the existing company's profits were soon obvious to everyone. They were too strong to be ignored.

Progress at Western Deeps in the early eighties was not without setbacks, however; success not without tragedy. On 31 October 1981 a winding accident — this time at Number 3 Shaft — claimed the lives of 18 men. The official enquiry revealed that the emergency braking system had failed, but that no blame attached to the mine or to any of its employees.

The continuing presence of vast quantities of underground water throughout the region could not be ignored. It was a nightmare which would not go away. Its existence impinged itself on the consciousness of planners, administrators and workers, and the dangers it posed could neither be overlooked nor lightly regarded. The drama of the flooding at West Driefontein, adjacent to Western Deeps, though some year previously, was still fresh in the memory.

Steps had been taken to eliminate the threat. The underground lakes which endangered mining operations in the area had been de-watered. At the new Number 1 Shaft, where shaft sinking was intensively in progress, standard precautions were taken and pre-cementation diamond-drilling probed the earth in advance of the shaft, which by mid-1982 was nearing the 1 000-metre mark.

Suddenly, on 9 July, a slab of about three metres in length and two in height broke away from the sidewall at a level of 970 metres, the base of the shaft at that time. Immediately water began to gush into the shaft.

Where did it come from? The geological compartment in which the shaft was being sunk had been de-watered. Why did the pre-cementation holes not locate the water? The total unexpectedness of the threatened flood shocked everyone. Every conceivable precaution had been taken.

Alarm bells rang. At once pumps were rushed into action. The water level continued to rise, however, and more pumps were brought in. The water went on winning. Hurriedly kibbles were used for bailing. Still the water rose, gaining at a rate of about one metre an hour. The pumps were removed and bailing was halted to enable the total rate of the inflow to be calculated. Without these counter-measures, the level climbed up the shaft at five metres an hour. Arithmetic quickly showed that from somewhere water was pouring into the shaft at 126 litres a second.

From experience the engineers knew that such a strong flow would rapidly erode the edges of the hole through which it was gushing and that the rate of flow would thus continue to increase.

Dick Solms, Project Mine Manager in charge of the work at the Number 1 Shaft, later pointed out that because "... the rate of inflow could thus increase until the pumping capacity was neutralised ... it was decided to allow the water to rise to its natural level ..." At that stage, without further inflow, it was felt, the sealing of the fissure could be more easily accomplished.

The decision was taken the morning after the flood began. The intruding water was left to find its own level. For five anxious days the water continued to creep higher and higher in the silent shaft, until it finally settled at 418 metres from the shaft collar, having risen more than 550 metres.

Consideration of the various optional methods available to tackle the salvage problem led to the decision to cast an underwater plug to block the hole and then to pump the water from the shaft before finally reinforcing the weak area in the shaft wall. Only then could shaft sinking proceed.

The Civil Engineering Department of Witwatersrand University, commissioned to design the plug, came up with an estimate of a plug 17 metres in length. To increase safety, engineers increased the length to 27 metres.

The reclamation work was not without difficulties. Pipes being lowered into the water broke away from the slings and, buckling as they impacted upon the shaft's bottom, settled "like a bunch of spaghetti". An engineer was flown out from the British North Sea Oil rigs to advise on the possible use of divers to inspect the damage. An underwater camera was used in an attempt to gain information about conditions deep in the shaft. In vain. It only showed that the murkiness made it impossible for divers to operate.

Endless setbacks hindered progress. Blinded by the intervening column of water, engineers had to proceed by "feel" — sensing their operations rather like sightless men reading braille, or, like people in darkness, making their way through a complex maze.

Some six weeks after the burst, the tipping of 4 300 tons of rocks began and cementation followed. Then, with the plug completed, dewatering commenced and the shaft was pumped out to the top of the enormous plug. Repairs continued and by degrees the damaged shaft was reclaimed. Only on 1 March the following year — almost eight months after the flooding began — did sinking operations get fully under way again.

It had been a costly affair, not least in terms of time lost. Reporting to shareholders, Gerald Langton said that all costs associated with the flooding and rehabilitation of the shaft were covered by insurance, but "this cover does not apply to the consequential loss . . ." In an attempt to offset the time lost, steps were immediately taken to accelerate the shaft sinking programme in an effort to reduce the consequences of disruption in the overall production plans.

For Western Deep Levels, gold profits (before tax) in 1980 had amounted to R582 885 000, more than double those of the previous year. Some years were to elapse before the figure was beaten. In 1980 Western Deeps paid a gigantic R200-million to its shareholders and more than R328,3-million to the State in taxation and as the State's share of profits.

A note in the company's annual report for 1982 tells that ". . . In two decades of production, revenue from 959 292 kilograms of gold has totalled R3 645,5-million. Taxation and the state's share of profit of R1 034,4 has been paid and capital expenditure has amounted to R679,1-million."

A golden egg, indeed!

The higher price of gold was not without certain adverse repercussions, however, as reactions in the market place quickly revealed.

Less and less is gold being used as a basis for currency. Other uses now predominate. A note in Western Deeps' 15th Annual Report (1972) mentioned that "World production of gold in 1971 totalled 1 250 metric tons, of which 1 058 tons, or 84,6 per cent, was used in the jewellery industry . . ." In the face of steeply rising prices, for the next few years the amount of gold purchased for use in jewellery declined. In his annual report for 1979, the Chairman said: ". . . There is no doubt that the current high prices are depressing the jewellery trade, partly because of consumer resistance but also because of the difficulty in financing working stocks . . ." The position changed sharply in 1981 when lower prices of gold led to a big upswing in purchases by the jewellery industry.

To promote the sale of gold to the jewellery industry and to market Krugerrands in Europe, Canada and the Far East, in 1971 The Chamber of Mines created the International Gold Corporation Limited — "Intergold". Marketing was later extended to America. In 1974 more than three-million of South Africa's production of 24,4-million ounces of gold were sold in the form of these coins.

The Chairman of Western Deeps said: ". . . Against a background of rampant inflation, a mounting energy crisis and an international monetary system in which there was little confidence, the movement into gold gathered momentum . . ."

Sales of Krugerrands increased further in the following year, and yet higher quantities of gold were absorbed by the jewellery industry. A record 4 800 000 Krugerrands were sold, despite the closing of the United Kingdom market and the threat — implemented in 1976 — by the German government to impose value added tax on all transactions involving gold coins. While sales dropped in the following two years, in 1978 the previous record was beaten by the sale of no less than six-million Krugerrands.

The value of marketing gold in the form of coins or medallions was quickly appreciated by other countries which soon followed suit, notably Canada and the United States. Now there are many more.

In 1976 the International Monetary Fund had begun a series of auction sales, implementing an agreement reached at a meeting held in Jamaica to sell a sixth of its gold holdings in order to create a fund to aid developing countries. It was aimed to hold these auctions at regular intervals over the ensuing four years. The first took place in June 1976. Other sales followed at intervals of six weeks with 780 000 ounces of gold being offered for sale on each occasion. The amounts were later reduced until such sales ended in 1980.

In 1978, however, the US Treasury began selling off some of its gold stocks in an effort to bolster the value of the dollar.

Inevitably these sales depressed the price of the metal.

The Chairman of Western Deeps, Dennis Etheredge, pointed out, however, that "... the downward pressure of IMF sales was being offset by the strong upward trend in industrial demand for gold which first emerged in the latter half of 1975. Indeed, it is thought that industrial off-take exceeded mine production in the Western World during 1976. Accompanying the growth in industrial demand, substantial sales were also taking place to the Middle and Far East, particularly to Turkey, Thailand and Indonesia. . . ." He went on to say that ". . . As confidence in gold has returned, so the impact of IMF sales has decreased. . . ."

For South Africa fluctuations in the world's gold price in the eighties has been offset by the depreciation of the rand against the dollar — the currency earned by gold.

Caught again in the scattered cross-fire of international politics, in 1987 Western Deeps created a new class of shares to enable United States nationals to continue investing in the company without breaching sanctions legislation adopted by the US against South Africa.

This issue of what is known as "S Ordinary" shares was created from 355 885 unissued ordinary shares of R2 each. To qualify as a legally distinctively separate class of share, the new issue, separate from any issued after 2 October 1987, carried an additional preferential right to the payment of one cent per share in the event of the company's being wound up.

If the celebration in 1986 of the centenary of the discovery of gold on the Witwatersrand and of the foundation of Johannesburg was less spectacular than its importance to the nation's economic well-being warranted, the industry itself had good reasons to be jubilant.

For the gold mines 1986 was "an exceptional year". The Chamber of Mines announced that "tonnage milled, revenues and profits all reached record levels". While the 638 tons of gold produced was almost five per cent less than that produced in 1985, working revenue was 17 per cent higher than that of the previous year. Profits were a record R8 307-million.

To the industry's achievements of that centenary year Western Deeps contributed substantially. Its own profits reached a record R574 551 000.

Its accomplishments placed it among South Africa's "Top 100 Companies for 1986". A certificate acknowledging the achievement is to be seen in the head offices of Anglo's Gold Division at 45 Main Street, where it hangs in the office of Chris Bull, Senior Secretary of the Gold Division dealing with Western Deeps.

However "exceptional" 1986 was seen to be for the gold industry, 1987 saw the average dollar price for the metal increase by 21 per cent to $447 US an ounce compared with $368 in the previous year, confirming continuing international confidence in the value of gold.

CHAPTER TWELVE

Quo Vadis . . .?

On the basis of early estimates of the working life of Western Deeps mine, in 1987 we now stand at the half-way mark, gazing back some thirty busy years and looking ahead over a similar span. Fresh evidence has led to a revision of those estimates, however, and the mine is now perceived as having forty more years of profitable existence — which takes operations to beyond the first quarter of the twenty-first century. And, who knows, as science and experience intertwine, as humans and machines collaborate and as more information comes to light, developments in the intervening years ahead may extend production yet further. Certainly operations will continue for as long as mining is feasible, however deep the gold lies, and as long as demand creates a viable economic climate.

The nature of the demand may change, as it has done before, the emphasis shifting, reposing, in varying degrees, sometimes in jewellery, at other times in money and investment. Through the millennia demand in some form has always remained, however, though inevitably levels have fluctuated.

There is now a new area in which gold is taking on an increasingly important role: industry. The sophisticated machines and delicate instruments of the coming technological age need gold, and there is a shift from ornamentation to utility in the metal's usage. In this field particularly gold's essential usefulness creates a special demand, which, if anything, will increase. Present high prices are making possible the development of numbers of low-grade sources of the metal, sources not viable when prices were lower, and for the moment fresh mines are adding to existing stocks. But will the new discoveries in many countries be substantial enough to meet the growing demands? And

Main Administrative Offices, Western Region. The artist has included the author (left) in this drawing!

for how long? And when, in the longer term, will supplies cease to meet growing demand for the precious metal?

Prophecy was ever a hazardous pastime. Predictions about the future production of gold are no less so. The factors involved are too variable, too uncertain: not only regarding the future of existing as well as new sources, but also in respect of the metal's value, which remains at the whim of an immense diversity of disparate factors.

Professor A.D. Pretorius, Director of the Economic Growth Research

Unit of the University of the Witwatersrand, has cast a gloomy pall over the future prospects for South Africa's gold industry. Since gold is the mainstay of South Africa's economy, his prognostications must impact on the country's future too. In an analysis (1981) of future supply and demand, he points out that production in South Africa, hitherto the leading supplier of the world's gold, declined by some 30 per cent in the seventies. In that same decade overall world production fell by 15 per cent, a decrease of 25 per cent in the Western World being offset by an increase of 15 per cent in the USSR in the same period. In those years, furthermore, the total demand for gold rose by 40 per cent.

In 1979 South African mines produced 703,5 tons of gold. In 1987 production had dropped 14 per cent to 605 tons. By the year 2000 South Africa's production is expected to drop to 50 per cent of what it was in 1979, unless (and there is always a proviso!) substantial new gold-fields are found in the country. By then, despite rising output from a number of western countries such as Canada, Australia, the USA and Brazil, increased production in China and the USSR will make the communist world the leading producer. Furthermore, by then, too, if prophesies are realised, the USSR will have displaced South Africa as the world's premier gold producer.

This, economists fear, will strengthen the USSR's ability to manipulate the amounts of gold available on the world's markets and thus to influence the price. The effects on the economies of western nations could be serious. For South Africa, with gold its major product, the impact would be heavy indeed.

The timing of this gloomy scenario falls well within the predicted lifetime of Western Deeps and, since the viability of production is wholly related to the prevailing price of gold, the outlook must be taken into account in estimating the mine's future. Operations must be finely honed to face the future years when increasing production costs coincide with diminishing returns.

The recommendations of the Margo Commission of Enquiry into Taxation, published during 1987, would appear not to help the gold mining industry of the future. The Commission recommended that the present formula tax be replaced by a flat rate of tax and that the existing system of immediate capital write-off be replaced by one in which capital is written off over a three-years period after assets have been brought into use. This system aims at replacing the capital allowance and capital write-off which, for many years, made the enormous risks of deep-level mining more acceptable to investors.

In his Chairman's review of 1987, Peter Gush points out that "... The adverse effect of a flat rate of tax and extended capital write-off, if they are implemented, should not be under-estimated. In the face of consistently reducing grades, the low-grade, low-margin mines will have to pay significantly more tax and many shafts could become vulnerable to closure ... Thus many current operations could be jeopardised and, in addition, investors could be discouraged from financing new mines because the risk profile of capital projects will have changed in favour of the fiscus at the expense of the providers of capital."

He further points out that any such closures will have a negative impact on communities dependent on such mines.

Concern goes beyond that. Any significant diminution in the profitability of the mining industry will affect not only certain local communities: any negative trend will influence the welfare of the country as a whole, and the well-being of all its people, in addition to that of the whole southern African region.

Furthermore, the problems arising from diminishing resources are likely to coincide with an anticipated burgeoning of the population.

Thought is being given to ways of softening the negative force in any slowing down in South Africa's mining industry.

It has been suggested that South Africa should channel more of its resources into its own industries; that the country should refine its own minerals, for example, and should use them essentially in local manufacturing, rather than exporting the materials in a raw and unprocessed state.

For gold this would mean establishing a large South African jewellery industry — exporting wrought gold instead of bullion — unfortunately located well distant from the main markets.

So much for the materials. What of the humans who will have to shoulder any burden arising from decreases in the prosperity of the mining industry?

South Africa is seen to have a number of options.

It has been suggested that further developing the country's immense tourist potential would create employment opportunities that will help to absorb large numbers of semi-skilled workers.

A comparison has been made between South Africa's gold mining industry and Spain's tourism. In 1986 that small country earned as much — about R7-billion — from tourism as South Africa did from its gold.

Western Deeps plans for the years ahead.

In conversation Gush asks how mining at depths below 4 000 metres

Underground workshop — for maintenance of trackless mining equipment.

can be carried out economically, and at once makes the question a rhetorical one by explaining how future operations will require new technology and fresh concepts. The magnitude of the developments and soaring development costs necessitate a totally new order of thinking: improvements cannot merely be slight modifications, he says.

The future of Western Deeps is planned around the new Number 1 Shaft. Twenty-four years or so younger than its elder brothers over the hill, Number 1 Shaft will take over as the other two complexes are phased out of operation. First, it is planned, Number 3 Shaft will be closed — "soon after the turn of the century," Fred Bayley says — and the final mining of pillars and residue in that complex will be done through Number 1 Shaft. Number 2 Shaft, with more ground, will continue probably until after the year 2010 before being shut down.

Operations at Number 1 Shaft may perhaps be supplemented by the addition of one further shaft complex. Consideration has been given to sinking a shaft on the south-east sector of the lease area, but current thinking would have it further south, even though the operating reefs there are much deeper — in excess of 4 000 metres and going down to 5 000. The values revealed by test drilling in the area will influence the final decision.

Geologists are now aided in their prospecting by a new generation of instruments and machines and by advancing techniques. One is the vibroseis technique. By sending sound waves underground and timing the return echo off rocks below, it produces a comprehensive profile of the underlying strata. This assists the geologist in determining the structure of the strata which enables the boreholes to be more appropriately sited. From the drilling results engineers gain the information that enables decisions to be made about positioning and sinking shafts as well as about the pattern of mining ahead.

Such seismic surveys could well form the basis of any drilling programme by Western Deeps to investigate those reefs lying at depths down to 5 000 metres south of the lease area and into ground now held by Western Ultra Deep Levels. Those possibilities, however, at present probably lie beyond the horizons of this century. In such an event, however, the techniques currently being pioneered at Number 1 Shaft will prove extremely useful for mining at those greater depths.

To meet the extremely hot conditions, for example, there will be new concepts in ventilation and cooling systems, just as wholly innovative support methods will be introduced to counter the high rock pressures. New techniques in design and layout will be crucial. Using state-of-the-art computer technology, the new design at Western Deeps will be totally integrated with cost factors to give maximum yields at optimum cost-effectiveness.

"Deeps will spearhead this exciting new mining approach," Peter Gush predicts, "evolving an integrated deep-level mining system which will become standard for the industry in South Africa."

These deeper ore deposits are too rich to be ignored: the depths at which they lie demand fresh techniques to mine them. The eyes of the mining world will therefore be closely focused upon operations at Western Deeps.

Of course much will depend on the ability of humans to work effectively at such depths. Taking men down to those levels on a daily basis could have adverse physiological effects and there could be some psychological impact too. The broad spectrum of human factors, physical and mental, will have to be taken into account and studied very carefully.

Theo Pretorius, Managing Director, realises that much has yet to be learned about mining at depth. He is realistic about the future of mining. "We don't know the full extent of the problems we are going to meet," he says: "in the circumstances we can really only guess." But he is confident that at Western Deeps "we will lick the problems as they arise." Western Deeps plans to develop a pilot winze and shaft to explore the problems at depth.

The planned extensions at Western Deeps have bred considerable confidence among planners and workers alike. The new plans, made subsequent to Professor Pretorius's analysis, may well cause changes to the implementation of the broad economic scenario he sketches for the coming decades.

Not only once has Western Deeps achieved what was thought to be impossible. For example, the mine is now operating at depths not long ago deemed to be beyond practical limits. Modestly Western Deeps claims to have begun where, in several instances, others have left off. In its totally integrated approach to exploration, its achievements have made the mine unique, while its impact on mining design and technology has been considerable.

Scientists and engineers are optimistic that a high percentage of all the gold encompassed by the Western Deeps lease area will ultimately be extracted. Economists are confident that the metal will retain its worth for all time to come. Future prospects thus look rosy.

Voicing this optimism, engineers from Western Deeps told a Mining and Metallurgical Congress in Singapore in 1986 that the Number 1 Shaft of the mine "will exploit the VCR and the Carbon Leader to 4 000 metres and will require a number of subshafts, thus allowing mining to continue to the year 2025." Such plans are already well advanced. The Sub-main shaft, extending 4 421,8 metres below datum, is already planned for completion in 2008. Explaining that the present lease area extends to below 4 000 metres below datum, they said that the company will take over further ground to the south of the existing lease where, in the twenty-first century, mining will move "below 4 000 metres towards 5 000 metres."

The term "datum" denotes an imaginary plane 1 828,79 metres above mean sea-level. It is a reference elevation that is a constant figure for use on all Anglo American mines.

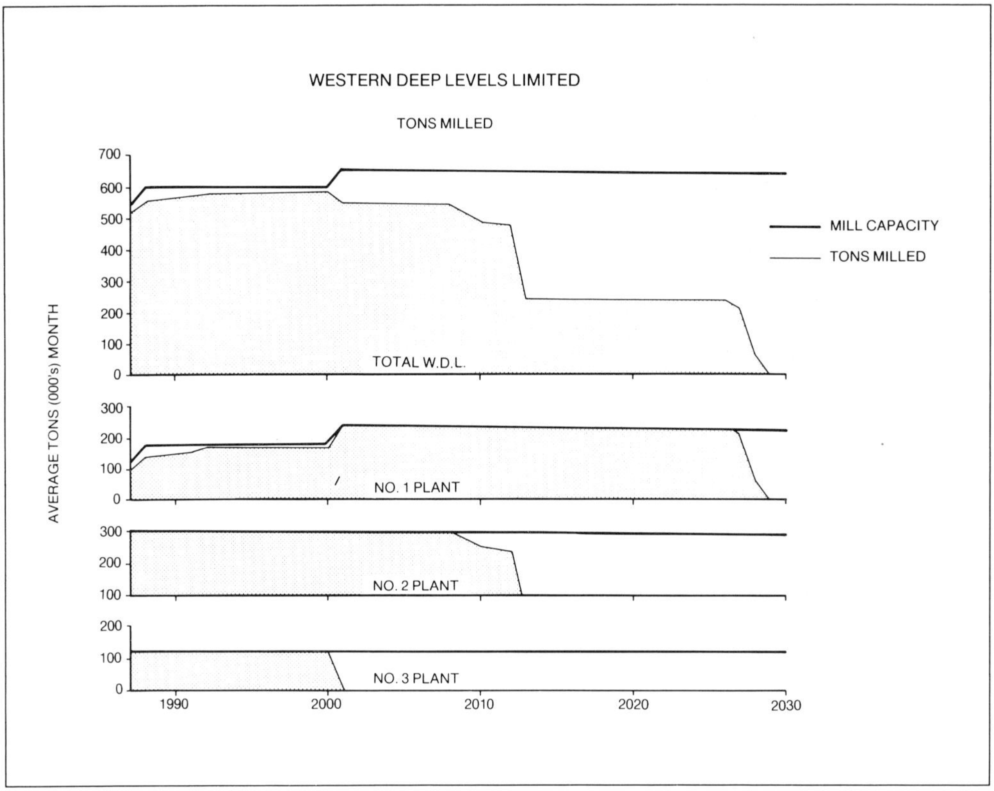

Tons milled and estimated production until 2030.

The sinking of shafts has become an extremely costly undertaking, in terms of both time and money. To counter at least some of the expenditure on further vertical shafts, Western Deeps has introduced a system of cross-cuts with high-speed haulage transportation that moves vast tonnages and which links with the vertical shaft systems. In an effort to increase the hoisting rate, thought is being given to equipping vertical shafts in future not with existing rope-hoisted skips, but with skips that operate on a continuous system. This will make it possible to handle greater quantities of ore in any given time. Consideration is being given also to re-examining the feasibility of hoisting mined rock in a hydraulic medium through pipes.

Present methods of using treated residue as backfill material mean bringing quantities of crushed rock back underground after treatment on the surface: costly both in terms of labour and power. To eliminate the need to return that bulk waste underground — having initially hoisted the rock in its raw state — consideration is being given to leaving the waste rock underground, keeping it for backfill. If one day it should prove feasible to crush the ore-bearing rock underground, only the gold would then have to be hoisted to the surface while the residue would immediately become available for use as backfill.

Gerald Langton sees this as one of the many advances which must essentially be made in mine development in the remaining years of the twentieth century. The economics and technology of such a plan have yet to be developed.

These developments, as exciting as they are, remain basically within the ambit of conventional mining practice. Essentially they are modifications, in some form, of existing technology. But wholly new — and different — methods, bordering on the realms of fantasy, are also being looked at.

Mind-boggling, some of these possibilities seem like creations of an overheated imagination, figments of wild, untrammelled science fiction. Yet, just as some of Jules Verne's fantasies of the last century have since come to present commonplace reality, so, for example, are laboratory experiments now edging the possibilities of mining with "bugs" — bacteria — closer to practicality.

"Perhaps the mine of the future lies in the bacteria," wrote a geologist at Western Deeps in 1987. "Imagine drilling into a gold deposit, fracturing it underground with small nuclear devices, and then sending an army of bacteria down the drillholes. These bacteria produce cyanide. The gold is dissolved by the cyanide, the resulting solution is then pumped to the surface and the gold is recovered in carbon columns. Another way: nature provides certain algae that undergo complex reactions with metallic ions. These could be sent down the "mine" after the bacteria, allowed to adsorb gold and then be pumped to the surface. The advantage of algae is that this habit of adsorbing gold continues after they are dead, so that they are 100 per cent cost effective."

On its front page on 18 September 1987, under the headline "Rock-munching bugs liberate gold", Johannesburg's *Business Day* described "a major breakthrough in extracting gold from refractory ores". Gencor's experiments in this field were described as giving South Africa "the first extraction plant in the world to use bacteria to eat rock — so releasing gold from certain types of ore." The article went on to say that "South Africa's plant is thought to be the largest in operation" and the world's first bioleaching operation to recover gold.

Refractory ores normally yield their gold with difficulty since they are mixed with arsenic and pyrite minerals, thus hampering the extraction of gold by conventional methods. The bacteria, *Thiobacillus ferro-*

oxidans, oxidise the pyrite into sulphuric acid and iron sulphate, freeing the gold.

Gencor's plant was reported to be treating some 300 tons of concentrates a month which, speculation suggested, could be producing as much as 45 kilograms of gold. Recovery in this manner is both cheaper and more effective than conventional methods, Gencor claims.

The newspaper went on to describe how, in developing these techniques, scientists in many parts of the world are using genetic engineering to improve the rock-eating habits of the bacteria and to develop strains able to tolerate high temperatures. The techniques open up new vistas for the future.

The commercial use of bio-oxidation in leaching minerals is far from being new, however. It was used in Spain more than 250 years ago to recover copper from mine dumps. For some years uranium has been recovered by bacterial leaching at certain gold mines in the Transvaal. Today its use is increasing in many parts of the world, as fast as laboratory experiments improve techniques and processes.

As land-locked deposits are exhausted by present industrial mining, the ocean floor gains in importance as an alternative source of minerals. On the Atlantic coastline of South Africa, De Beers mines diamonds, while off the Indian Ocean shores of the Cape petroleum gas is being sought: manganese nodules lie — plentifully — in the deep trenches of the Pacific. The range of available minerals in the sea is a wide one. Not only does the sea-bed have large deposits of many minerals, but the waters of the ocean also contain soluble concentrations of many mineral elements. It has been said that although the concentrations are very low, it is easier to treat large quantities of water than large masses of solids from the earth's crust.

Traces of gold are also to be found in the waters of the oceans. Quantities are not significant, however. But, depending on the economics of extraction, one day the sea might provide a further source of gold.

But whatever exciting and stimulating possibilities these developments might hold for gold-mining — and therefore Western Deeps too — in the years ahead, there is little doubt that the really great challenges during the coming decades will arise from the developing social and political changes which will take place in South Africa in the evolution of the post-apartheid era.

Anglo American has long been in the forefront of such advance thinking and in anticipating such changes and in endeavouring to stimulate and accelerate progress in all spheres of South African society.

As a part of the Group, Western Deeps has shared in this drive for social progress. Something of that vision has been made manifest, for example, in the establishment of Wedela town and of the Technical High School, respectively the firsts of their kind in the country.

Western Deeps led the mining industry in developing a system of participative management, in giving the residents a voice in the administration of the hostels, in improving their living conditions, and so on. At first this was unique to Western Deeps. It has since been instituted elsewhere.

This forward thinking flowed through to the work system, and Western Deeps was the first mine to set up "Safety Committees", bodies on which the workers themselves had their own representation. Through participation in "value circles", workers there became more involved in management than elsewhere, resulting in an increase in work-motivation and an enhanced sense of responsibility.

Western Deeps' approach to its people is enlightened without being paternalistic. The nature of mining operations makes collaborative team-work essential, and the company has always been very "people-conscious". Peter Gush tells that "Deeps" is a tough mine to work in and people are therefore exceptionally conscious of the involvement of co-workers in the total operation.

Sam Moyo, currently Assistant Public Relations Officer for Anglo's west region, explains how the labour situation at Western Deeps has changed over recent years. Whereas for a number of years black mine workers tended to be migratory, now they tend to stay for a long while, enjoying the facilities provided at the mine and benefiting from the security which conditions provide.

He himself has been employed by Anglo for twenty-four years, twenty-one of which have been at Western Deeps. He is looking forward in 1989 to receiving a gold watch which goes to all employees in the Corporation who have served continuously for twenty-five years.

For example, at one such presentation ceremony at Western Deeps in December 1987, no less than thirty-eight blacks were among the recipients of long-service awards, two of them with the special award for forty years of service. Moyo tells that there are now several blacks at Western Deeps with forty years of service to Anglo American — and even a sprinkling with fifty years! Service is specially recognised by the Corporation (and De Beers) with rewards and mementoes at twenty-five years and again after forty and fifty years of service. In each of the past five years, between sixty and a hundred employees — black and white — on Anglo's gold mines have received long-service awards.

Consistent with its concept of social responsibility, in September 1987 Anglo American announced a scheme to assist most black South African workers on its mines to acquire their own houses. Other mining groups are following suit. Special loans are available and repayments will be subsidised by the companies, no house-owner having to pay more than 25 per cent of his income. Whereas previously family housing was available only to a small number of senior, skilled mine employees, this scheme will enable more workers to bring their families to live with them near the mines. Other workers, presently unable to afford houses, are availing themselves of loans to buy plots on which to build homes in due course.

For those who, for any reasons, do not wish to bring their families — or who cannot bring them — the mines plan to continue with the hostel system and will thus provide a choice of accommodation for these single, unaccompanied residents.

It is hoped that the government will assist in the development of these towns and housing centres by providing schools and hospitals as well as other infrastructural services.

The families of foreign migrant workers are precluded from accompanying the bread-winner to South Africa, however. Since foreign workers comprise 41 per cent of the work force on Anglo's gold mines, large numbers will be compelled to continue living in hostels. Such accommodation is being continually improved.

To give employees of Anglo American a personal and direct interest in the Corporation, in October 1987 those with at least two years' service were given an opportunity of becoming shareholders by being allocated five shares each per year, the shares being paid for by the Corporation. The scheme was conceived some years earlier and worked out in detail after considerable research into share ownership schemes in several countries.

In his annual statement to the Corporation's shareholders in July, the Chairman, Mr Gavin Relly, had said, "... In South Africa, where the wealth-creating process of the First World must arrive at a durable synthesis with the needs and aspirations of the Third, there is a strong case for believing that the stake held in the country through growing home ownership can well be matched by workers holding a direct stake in the business in which they are employed." He went on: "This view is surely consistent with the world trend away from centralist socialism on the one hand and rigorous capitalism on the other, to something in-between, founded not on ideology but on pragmatism, deriving its strength and support from the fact that it is seen to work."

This Group Employee Shareholder Scheme was put into practice early in 1988 when all 2 445 employees at Head Office with at least two years' service were each offered five shares in the Corporation. Of these, 2 402 accepted: only forty-three declined.

Associated companies of the Corporation were invited to join the scheme and, if all accept, a further 250 000 employees qualifying for participation in the scheme will similarly each be allocated shares, the number offered being decided by the respective companies. Western Deeps decided to adopt the scheme and that decision was implemented in March 1988.

In a message addressed to employees of the Corporation eligible for participation in the scheme, Mr Relly said, ". . . We believe it important for employees to have the opportunity to own an interest in the Corporation and in consequence to involve themselves more fully in the processes by which a free enterprise system works."

The scheme will be reviewed after five years. In the interim, further offers of shares will be made annually to employees who qualify for participation. Shares allocated will be held by a trust fund for a period of four years, during which participants may exercise their voting rights by giving written instructions to the trustees. At the conclusion of that period, employees can choose whether to leave their shares with the trust, to take delivery of them, or sell them. The scheme will enable its beneficiaries to appreciate the merits of capitalism. Participation will give each a stake in the organisation of which he is a part, enhancing the individual sense of responsibility.

The cost to the Corporation over the initial five-years period will run to many millions of rand, the amount depending upon the share prices of the participating companies. High levels of individual participation will mean that some 3,5 per cent of Anglo's present share issue capital will be in the hands of its employees.

The financial benefit of the scheme to the employee will depend in large measure upon the success of the operations in each company. Such success depends, in turn, upon the totality of the contribution made by workers at all levels and in every sector of operations. In some measure this shifts the consequences of any strike action from the company to the individual employee who now, with a stake in the mine, more than ever stands to lose. There are critics who point out, however, that since dividends do not match wage increases such a belief has no real foundation: more could be gained from militant labour action than from the benefits of capitalist participation.

Speaking of the consequences of the severe strike by mine workers in August 1987, Mr Relly pointed out that the strike imposed a "real economic burden on workers," as well as on companies and the entire economy of South Africa.

In the longer run the economic consequences of strikes clearly put jobs at peril, especially in the face of the ready alternative in the form of mechanisation. Also referring to that strike, the Chairman of Gold Fields of South Africa pointed out that excessive and frequent wage demands would lead to the closure of mine areas and the consequent reduction in job opportunities. Furthermore, the economics of such demands would compel employers to speed up mechanisation — with consequent reductions in the availability of jobs. In similar vein, the Chairman of Western Deeps pointed out in his Review of 1987 that "experience gained during the strike has led to a programme of rationalising mine operations using less labour".

The National Union of Mineworkers (NUM) claimed that the strike cost mine owners R250-million. But the consequences were more than that. Naas Steenkamp, a director of Gencor, who only a few weeks earlier had succeeded Peter Gush as President of the Chamber of Mines, spoke of the costs which had also to "be measured in terms of the strains imposed on the management-worker relationship, the strife and tension between strikers and non-strikers, the disturbing incidence of violence, intimidation and murder, the loss of jobs and the wages sacrificed." NUM's decision to call off the strike was welcomed not only by worker and management, but also by the country as a whole. It had gone on longer than most workers had wished and, there was little doubt, had been continued under political pressure from outside the country.

In this case NUM, which had earlier broken off negotiations, eventually settled for the wage increases offered by management before the strike. NUM had called for a 60 per cent increase in wages, while the industry had offered between 15 and 23,4 per cent, depending on the task category. Refusing to admit defeat, however, the Union described its prolonged walk-out as "a dress rehearsal for further action".

It is to be hoped that any further action will not be hastily undertaken nor implemented without responsible consideration of the broadest consequences. The country has sufficient economic and political problems without any being added. If credibility is given to the concept that from outside its borders South Africa is being subjected to a "total onslaught", then it must be accepted that the gold-mining industry — mainstay of the nation's economy — must rank as a prime target, with the more productive mines, such as Western Deeps, a bull's-eye within that broader target. The possibility is as dreadful as the rationale may be individually flattering. *Terrorists don't hijack paupers!*

The world must come to appreciate that, since the rest of southern Africa is economically closely linked with the Republic, to harm this country is to wound its neighbours and, as the ripples spread out, to reduce the influence and security of western representation and authority in the region.

There is probably no better illustration of the importance of South African mining to the southern African region than that which is made plain by the amounts of money paid by South Africa's mining industry to other African states in terms of labour agreements required by their governments. In 1985, for example, some R1 041,7-million in compulsory deferred pay and remittances was estimated to have been paid to these states from the pay packets of their nationals working in South Africa. Half Lesotho's male labour force is employed in South Africa and remittances of pay account for almost half that country's GNP.

The massive contribution made by the mining industry to South Africa's own economy needs no further emphasis. Given the many-faceted role which gold plays in world affairs and the size of South Africa's contribution to the world's gold treasury, the industry in South Africa has an importance not only to itself and its workers, nor only to the nation, but also across a far wider international economic and political spectrum. This importance, generally well appreciated by the large mining corporations, deserves to be appreciated equally responsibly by the workers and their trade unions — as well as by responsible politicians of the Free World.

Black trade unions in South Africa are not a new phenomenon. They have existed for sixty years. Their history is not altogether a proud one, nor, similarly, can that history be a matter of pride to some governments which have held sway in those years.

Discussing the matter in *Optima* in December 1983, Professor Roux van der Merwe of the University of Port Elizabeth points out that "The National Party won the general election of 1948 in part because of the objection of white mineworkers to the governing United Party's efforts to bring the growing black industrial force within the ambit of statutory labour structures."

All that is now past, however. Legislation has since given recognition to registered trade unions and the benevolent liberalism of industrial management, now less fettered by law and regulation, is endeavouring to improve the lot of the black worker. The need, now and in the future, is for greater co-operation and less confrontation. Both management and the unions need each other, "both parties willy-nilly are part of the free-market system . . ." and Professor van der Merwe goes on to say: as unpalatable as the unions may find it, ". . . the capitalist system is at one and the same time a part of their *raison d'être* and their guarantee of survival as champions of the workers' cause."

Peter Gush has pointed out how in recent years the NUM has sacrificed the interests of miners in favour of direct political goals. Calls for disinvestment and sanctions and for the nationalisation of the mines have been increasingly advanced. Such attitudes can currently only be detrimental to the industry, and thus to the future of the unions. Professor van der Merwe points out that "Without exception, in post-revolutionary societies the union movement has lost its place as the defender of working-class rights and interests. Nowhere is this better demonstrated than in post-colonial Africa. In the 1950s and 1960s many Africans nursed the hope and vision of a non-racial and democratic world wherein their fecund continent could move towards its almost limitless potential. . . . The reality has been catastrophically different. . . ." In most cases trade unions have been subordinated to the state.

Eric Brereton, a former General Manager at Western Deeps and a recently retired Chairman of Shaft Sinkers (Pty), regards South African miners as the best in the world. Much travelled in the course of his business career, he has got to know the international mining scene and has developed a high regard for people in the mining industry everywhere. But he knows of none better than those — of all races — in South Africa where, he says, their dedication to their work is head and shoulders above those attitudes and attributes elsewhere in the world.

Over the years he has noticed the leftward shift to socialism in mining industries overseas, and has seen how the unions are claiming more and more attention, to the detriment of the industry itself and with self-defeating personal consequences. Africa's future cannot be complacently regarded. Benefits brought long ago from other lands reduced the continent's death rate substantially — but not the birth rate. In many African lands, domestically generated subsistence resources have become inadequate for the burgeoning population and famines have consequently become widespread throughout the continent.

For South Africa, however, its natural wealth and its skills currently keep such tragedy at bay. But signs for the future are ominous. Already the population growth augurs hardship in the years ahead. Government surveys indicate that with the present rate of growth, South Africa's population will have almost doubled to 47-million people by the end of this century. Even with an economic growth rate of three per cent per annum, it is estimated that eight-million South Africans will be without employment by the year 2000.

Education thus becomes vitally important. The development of skills which can contribute to the nation's welfare has become a matter of urgency. The need to learn takes on a desperate quality. In its own domain Western Deeps has provided the facilities.

Commenting on the effects of the miners' strike in August 1987, Theo Pretorius, Managing Director of Western Deeps, pointed out how the consequent loss of profits had resulted in the cancellation of projects beneficial to employees and a resultant loss of some jobs. In a world facing increasing unemployment, he said, "I believe having a job is going to be one of the most precious things in the future."

South African miners and technicians are knowledgeable, each in his own sphere: they are well led and well motivated. Their loyalty and industriousness are absorbed into the spirit of the mine which becomes further apparent in the quality of the industry as a whole.

"It is those special qualities which have made Western Deeps great," he states — without fear of contradiction. The particular technical and logistic problems that are a part and parcel of Western Deeps make work there tough. That toughness sorts out the men from the boys. Only those able to stand the problems survive: others leave. Those who remain are special people with special qualities. Soon they knit into a close team that does superhuman things when required.

It is not only the depth of the mine which has made Western Deeps known all over the world, nor merely its advanced technology. Those targets would not have been achieved without dedicated people. The extent to which visitors from all over the world come to see Western Deeps is an indication of its international renown and a tacit tribute to that dedication.

A Visitors' Book at the Number 2 Shaft smelthouse is a treasure chest of names of persons who, through the years, have visited Western Deeps; people who have come in their hundreds not only from all over South Africa and many walks of life, but also from foreign capitals as well as remote corners of the world. Old and young, they span a fascinating range: Cabinet Ministers, servants of governments, representatives of companies, private individuals, scientists and miners, technicians and businessmen; people from all occupations. All have left signatures of varying characteristics. Among addresses from every-

April 1981: Minister of Mines, the Hon F. W. de Klerk, visits "the deepest point in the world": the bottom of Number 2 tertiary shaft.
Left to right: *Fred Bayley, Consulting Engineer; Dennis Etheredge, former Chairman of WDL; Minister de Klerk; Eric Brereton, General Manager, WDL.*

where are entries such as "Bank of England," "Royal School of Mines", "House of Commons", "US Congress", "IAEA, Vienna". Western Deeps has hosted them all. The Deputy Prime Minister of the Netherlands, Dr J.A. Bakker, accompanied by other personalities of the Dutch Government, in November 1970 went underground at Number 3 Shaft, his wife upset that women were discouraged from going too.

The main target for visitors is the bottom of a tertiary shaft at the Number 2 complex where, in the usual darkness, there is now a sign which tells that this is the lowest point on earth. It is 3 581,7 metres underground and 1 949 metres below sea level. It will not be long at Western Deeps before that sign is out of date there. It will have to be moved — and, doubtless, through the years will be moved again and yet again.

Though it will doubtless one day be challenged by a new generation of developing gold mines, for the remainder of its existence, Western Deeps seems destined to continue to lead the world in deep-level mining; generating further "firsts" as it goes, adding to the long list of significant achievements it has already compiled.

Built on sound foundations by men of vision and dedication, Western Deeps today reaches into the future with the same inspiration and faith. Generations of today, heirs to those enduring qualities, are dedicated to similar achievement and are filled with a desire by example to create traditions which will further enhance the reputation of the mine they serve and, in turn, imbue their successors — tomorrow's workers — with similar ambitions, respect and veneration.

The spirit of great men haunts its past. The roll-call of its founders echoes through the history of South Africa's mining industry.

First among them, Sir Ernest Oppenheimer, whose acumen, vision and courageous foresight is exemplified in the mine he created. His instinct led to the choice of great geologists such as Bancroft and Unger, inspired the co-operation of financiers and administrators of the calibre of R.B. Hagart and Bill Wilson, of engineers such as Harry McConachie — and many others, on and on down through the decades. To their dedication and ability Western Deeps will always be a vibrant tribute.

Something of that inspiring faith emerged again, a quarter of a century later, in leaders such as Gerald Langton and Rawdon Lawrie whose initiative and courage led to the next major development at Western Deeps: the major extension in the form of the great Number 1 Shaft, born of past experience, with its feet in the present and its eyes upon the future.

Through the years, those early schemes and dreams have been developed to reality by thousands of dedicated workers, both black and white, women as well as men.

Some thirty years have seen the long and distinguished register of workers at Western Deeps grow: men of Africa, men from other parts of the world. All have contributed to its greatness, their names written on gold.

Today more than ever the magnitude of Western Deeps makes it a mine where developments cannot be small. Its greatness dictates the scale of its operations: the power of its momentum, developed from the past, carries it high into the future.

From its inception characterised by innovation, in many of its operations Western Deeps began where others left off. Fresh goals lead it forward, plans leaping today's barriers to conquer and make realities of the technological challenges of tomorrow.

It is the deepest mine in the world: the deepest in history, reaching down through rocks compacted over thousands of millions of years, transitting strata that were ancient even before life on earth began: probing the Stygian darkness, sensing the heat of Hell itself: men with the courage and the hearts of lions working down where no lion ever walked.

Appendices

Appendix	A	Chairmen.
	B	Managing Directors.
	C	Directors.
	D	Chairmen of Western Deep Levels/Presidents of Chamber of Mines.
	E	Chamber of Mines Bravery Awards: Western Deep Levels.
Appendix	1	Gold produced (by years): Annual profits and tax.
	2	Uranium produced (by years): Annual profits.
	3	WDL share prices: Johannesburg Stock Exchange.
	4	Labour force.
	5	Conversion table.

APPENDIX A

WESTERN DEEP LEVELS LIMITED
Chairmen: 1957 – 1987

Sir Ernest Oppenheimer	9 August 1957 – 25 November 1957
H. F. Oppenheimer	January 1958 – February 1963
H. C. Koch	February 1963 – July 1969
J. W. Shilling	August 1969 – April 1974
D. A. Etheredge O.B.E.	May 1974 – December 1977
G. Langton	January 1978 – June 1983
E. P. Gush	June 1983 –

APPENDIX B

WESTERN DEEP LEVELS LIMITED
Managing Directors: 1957 – 1987

J. W. Shilling	September 1957 – December 1962
H. C. Koch	January 1963
D. B. Hoffe	February 1963 – September 1964
W. R. Lawrie	August 1974 – March 1985
T. L. Pretorius	April 1985 –

APPENDIX C

WESTERN DEEP LEVELS LIMITED
Directors: 1957 - 1987

Julian Amery: M.P.		G. Langton	1978 -
The Rt Hon. H	1965 - 1970	R. S. Lawrence	1971 - 1984
	1975 -	W. R. Lawrie	1973 - 1985
A. D. Backwell	1971 - 1972	N. W. S. Lewin	1957 - 1963
H. C. Ballingall	1963 - 1968	A. Louw	1964 - 1981
A. I. Beck	1968 - 1973	D. E.MacIver	1974 - 1975
F. S. Berning	1964 - 1969	H. G. McKerrow	1957 - 1958
L. D. Browne O.B.E.	1960	J. B. Mudd M.B.E., M.C.	1963 - 1968
W. J. Busschau	1958 - 1964	T. F. Muller	1969 - 1971
M. W. Dunningham	1974 - 1975	A. H. Munro	1984 - 1986
J. G. Edmeston	1973 - 1974	C. R. Netscher	1984 - 1987
D. A. Etheredge O.B.E.	1971 - 1984	S. C. Newman	1968 - 1971
J. M. M. Ewing	1960 - 1965	M. C. O'Dowd	1970 - 1971
C. T. Fenton	1981 -	N. C. Officer	1987 -
R. J. J. Fourie	1979 - 1981	Sir Ernest Oppenheimer	1957
W. M. Frames	1957 - 1958	H. F. Oppenheimer	1958 - 1979
P. L. Grimbeek	1978 - 1984	A. C. Petersen	1971 - 1979
E. P. Gush	1983 -	R. A. Plumbridge	1969 - 1981
R. B. Hagart	1957	T. L. Pretorius	1985 -
E. S. Hallet	1957 - 1958	T. Reekie	1957
P. S. Hammond	1957 - 1960		1958 - 1971
A. R. Harrison O.B.E.	1959 - 1963	A. W. S. Schumann	1971 - 1978
D. B. Hoffe	1963 - 1964	J. W. Shilling	1957 - 1965
	1975 - 1977		1969 - 1974
N. A. Honnet	1981 - 1983	H. R. Slack	1979 - 1983
R. A. Hope	1966 - 1969	J. Ogilvie Thompson	1984 -
P. R. Janisch	1981 - 1984	D. T. Watt	1984 -
C. G. Knobbs	1983 -	W. D. Wilson	1957 - 1963
H. C. Koch	1962 - 1969	A. J. Wright	1986 -

APPENDIX D

CHAIRMEN OF W.D.L.
who have been
PRESIDENTS
of the
CHAMBER OF MINES

H. C. Koch	1957 – 1958	J. W. Shilling	1971 – 1972
	1961 – 1962	D. A. Etheredge	1979 – 1980
	1965 – 1966	E. P. Gush	1986 – 1987

APPENDIX E

CHAMBER OF MINES AWARDS FOR BRAVERY
Western Deep Levels

George Jagals	1969	Dr Brian Parker	1973
Charles Palm	1971	C.L. de Jager	1973
Paul van den Heever	1971	Rudolph Vorster	1987

CHAMBER OF MINES AWARDS FOR MERITORIOUS CONDUCT
Western Deep Levels

Rudolph D.J. Opperman	1987	Paul Desmond Haarhoff	1987

APPENDIX 1

GOLD PRODUCTION, ANNUAL PROFITS AND TAX

	Tons milled 000s	*Yield dwt per ton*	*Working Profit R000s*	*Taxation R000s*
1962	1 400	6.89	4 604	–
1963	2 275	7.06	7 832	–
1964	2 409	7.06	8 945	–
1965	2 742	8.51	14 076	–
1966	2 965	9.99	19 005	–
1967	3 088	10.16	19 785	–
1968	3 384	11.08	25 856	–
		grams/ ton		
1969	3 115	20.46	31 802	–
1970	3 171	19.05	25 869	–
1971	3 174	18.51	27 543	–
1972	3 140	19.20	45 607	9 595
1973	3 209	18.29	84 249	44 487
1974	3 020	15.93	114 138	64 054
1975	3 095	15.42	111 811	58 549
1976	2 941	15.14	76 366	38 126
1977	2 977	14.61	88 930	45 054
1978	3 223	14.17	145 590	75 492
1979	3 241	14.78	280 952	136 001
1980	3 329	13.70	582 885	328 303
1981	3 156	12.36	331 630	126 360
1982	3 499	11.28	304 206	108 329
1983	3 522	11.20	362 188	139 917
1984	3 576	10.08	361 413	73 155
1985	4 076	9.19	526 660	164 408
1986	5 527	6.73	574 551	207 074
1987	5 826	5.89	441 200	153 200

(Source: WDL Annual Reports)

APPENDIX 2

URANIUM PRODUCTION: ANNUAL PRODUCTION AND PROFITS

	Tons treated	*Yield kg/ton*	*Uranium Oxide produced. tons*	*Profits R000*
1972	616	0.26		7
1973	620	0.25		6
1974	616	0.22	136	1 328
1975	632	0.23	142	661
1976	603	0.23	140	1 719
1977	747	0.22	167	3 249
1978	874	0.21	183	2 341
1979	906	0.22	199	9 484
1980	1 629	0.13	213	3 501
1981	2 331	0.09	212	5 072
1982	2 272	0.08	183	2 214
1983	1 880	0.09	174	3 509
1984	1 606	0.09	146	1 567
1985	623	0.09	54	12 248
1986				18 138
1987				37 300

Source: WDL Annual Reports

APPENDIX 3

SHARE PRICES: WESTERN DEEP LEVELS
JOHANNESBURG STOCK EXCHANGE

	Low	*High*
	(cents)	
1968	920	1 310
1969	750	1 150
1970	690	870
1971	565	875
1972	645	1 100
1973	910	2 000
1974	2 000	3 600
1975	1 375	2 885
1976	850	1 800
1977	940	1 390
1978	1 075	1 700
1979	1 360	4 075
1980	3 400	8 825
1981	3 850	7 400
1982	2 350	7 950
1983	4 800	8 450
1984	5 675	8 400
1985	6 250	11 150
1986	9 900	19 500
1987	12 200	23 750

(Source: JSE Handbook)

APPENDIX 4

LABOUR FORCE: WESTERN DEEP LEVELS

	White	Black
1957	N/A	
1958	N/A	
1959	262	2 394
1960	328	2 622
1961	714	5 851
1962	983	8 229
1963	1 227	10 527
1964	1 413	11 279
1965	1 515	12 618
1966	1 564	13 227
1967	1 494	13 234
1968	1 478	13 731
1970	1 556	14 720
1971	1 597	15 227
1972	1 605	15 248
1973	1 577	15 861
1974	1 597	15 122
1975	1 645	13 586
1976	1 690	14 264

	Surface	*U/ground*	*Surface*	*U/ground*	*Total*
1977	593	1 267	2 606	13 778	18 244
1978	600	1 290	2 674	13 942	18 506
1979	603	1 270	2 785	13 404	18 062

	Surface	*Underground* *(All races)*	*Total*
1980	3 571	14 507	18 078
1981	4 768	14 990	19 758
1982	4 100	15 056	19 156
1983	4 123	15 275	19 398
1984	5 211	16 457	21 668
1985	6 081	18 984	25 065
1986	6 905	20 647	27 552
1987	6 794	20 671	27 465

(Source: Annual Reports: Western Deeps Levels)

APPENDIX 5

CONVERSION TABLE

(metric units to imperial units)

1 metre (m)	=	3,2808 feet
1 centimetre (cm)	=	0,394 inch
1 hectare (ha)	=	1,1675 morgen
1 centare (ca)	=	0,2990 square fathom
1 metric ton (t)	=	1,1023 short tons
	=	0,9842 long ton
1 gram (g)	=	0,6430 pennyweight (dwt)
1 kilogram (kg)	=	32,151 ounces
1 gram per metric ton (g/t)	=	0,583 pennyweight per ton
1 centimetre-gram (cm-g)	=	0,230 inch-pennyweight
1 kilogram per metric ton (kg/t)	=	2,0 pounds per ton
1 centimetre-kilogram (cm-kg)	=	0,7874 pounds per ton
1 litre (l)	=	0,220 gallons
1 megalitre (Ml)	=	219 969 gallons

Source: WDL Annual Report: 1970.

In 1961 in South Africa the pound was converted to R2. R1 = 100 cents.

South Africa pioneered the decimalisation of sterling. Australia and New Zealand followed suit some years later on the South African pattern. The United Kingdom did likewise some years afterwards, using the pound as a base, however.

Internationally the gold price is quoted in US dollars per fine ounce.

Acknowledgements

The multi-faceted nature of this fascinating subject has required researches over a broad field and I have thus had to range widely in search of information, not only through files and publications but, more importantly, through contact with many people both at the mine itself and elsewhere.

The list of collaborators is a long one. Their help has been invaluable. But for every one I mention, there will be, I fear, others left out. But though here perhaps unsung, they too have helped and I remain indebted to them. Through the years, with the growth of the workforce, individuals have tended to become absorbed into teams.

No account of Western Deep Levels — nor, indeed, of South Africa's mining industry — would be complete without mention of Mr Harry Oppenheimer, who has watched over Western Deeps since its inception and who, more than any other, has witnessed the development of the country's gold and diamond mining industries. I am most grateful to him for his kindness in contributing the Foreword to this narrative.

I was privileged also to speak with other Chairmen past and present whose advice and personal assistance proved very useful. I thank Peter Gush and Gerald Langton for the readiness with which they contributed such help. A few weeks after I had spent an informative afternoon with Dennis Etheredge, we were saddened by his death. John Shilling's long association with Western Deeps — as the first Managing Director and then as Chairman — and the breadth of his links with the mining industry added perspective to his views. This proved of considerable assistance in assembling this account of Western Deeps, and I share in the readers' indebtedness to him for providing much background about its early years.

Information, both spoken and written, from various generations of engineers, administrators, geologists, metallurgists and other experts, has ensured coverage of the history and progress of Western Deeps through the thirty-odd years of its existence. Managing Directors during those years — Douglas Hoffe and Rawdon Lawrie — shared their views and recollections, and to them I am grateful; as I am especially also to Theo Pretorius, the present MD, who, besides providing useful guidance and information, has been helpful throughout the preparation of this story.

Official documents provide useful information, but, generally laconic and limited to hard facts, often lack the human touch, without which any account becomes coldly impersonal, and Western Deeps is far from being an impersonal mine. The very toughness of its operations develops close team spirit and breeds an unusual camaraderie among those facing its challenges. Contact with men whose total service spans the lifetime of Western Deeps provided useful information — often more than could be used — which breathed life into what would otherwise be a stark narrative. To them I am immensely grateful.

Among these I would name Barry Mudd, who devoted several days to talking about the mine's inception and growth — with which he was closely associated — and, more than that, later kindly gave further time to checking technical details in the manuscript. Of his early contemporaries at Western Deeps, I had useful discussions with Dick Lambert, Andrew Tennent, Walter Wise, Dick Davidson and Tony Gyngell. The continuing success of Western Deeps is due, not least, to the soundness of the foundations laid by those early generations of dedicated technical and administrative personalities.

The tradition of excellence continues, passed on by men such as Eric Brereton, Fred Bayley, Frank Jantzon, Andries Vellema, David Dutton, Theo Meyer and Neels Moller, all of whom kindly gave time to talk about their associations with the mine. There are surely many others — associated at various times with Western Deeps — whom circumstances put beyond my reach. For that the narrative must undoubtedly and regrettably be the poorer.

I was fortunate enough to visit the mine a number of times. Always I was received with great courtesy and help and generous hospitality. This friendliness and co-operation began with Mr Fred Bayley and his wife and extended to all whom I there encountered. I shall long remember with gratitude their kindness. Colin Latulipe, Catering Manager, and his staff helped to make my visits comfortable.

In discussions about the geology of the region, I enjoyed Hugh Elliot's accounts of the formation of our world and the deposition of gold in the Witwatersrand Basin. I found equally rewarding my discussions with the scientists handling Rock Mechanics at the mine, Dave Arnold, Terry Hagan and Van Zyl Brink, who, with their colleagues, are impressively dedicated to making underground workings safer.

The mine's hospital is worthy of the pride with which it is regarded. Medical treatment of black workers is often caught between advancing science, on the one hand, and primitive custom and beliefs on the other. I am grateful to Pat Gillies, Medical Superintendent at WDL, for giving time to discuss the needs, present and future, of health care and social responsibility in an African mining environment.

Humphrey Hobday, Manager of Number 1 Shaft, was generous with his time and with information about the role of the new complex and, with Rob Willemse, Assistant Manager, kindly arranged for me to see gold poured.

David Hamilton, in charge of Strategic Planning at Western Deeps, and Derek Lloyd made themselves available to answer questions on technical matters, while Danie Jordaan, Survey Manager, not only provided detailed information about the mine property, but also took me on a tour of landmarks in the region. Peter King, head of Administration and Finance at the mine, gave me valuable background information about its management and its operations. Alton Williams, Manager, recounted interesting events during his long service at Western Deeps. Martin de Beer, Public Relations Officer, and Sam Moyo, his Assistant, were similarly helpful with information about people and events. Terry Gibson efficiently helped to provide diagrams to illustrate the text.

With modesty and a total lack of heroics, Corrie de Jager described the drama in which he was involved attempting to save the life of a Basuto miner trapped by a rockfall underground. Deservedly it earned him — and Dr Brian Parker — a Chamber of Mines Award for Bravery. Tales of similar valour came to me from written records.

Behind every successful man, it is said, there is an efficient secretary. In that fashion contributing to the success of the men they loyally serve are several at Western Deeps whose assistance to me proved valuable. In particular, for such help I remain indebted especially to Dawn Hanger, Annetjie Woods, Christine Watson and petite Heather Smit, with Springbok colours for gymnastics, whose feet during the week are firmly on the ground at Number 1 Shaft but hang suspended from a parachute at weekends as, skydiving, she increases her tally of sporting para drops.

At Head Office, I received considerable advice and help from Chris Bull, Senior Divisional Secretary, who handled my frequent queries with patience. Kent McNamara, now Industrial Relations Consultant to the Gold Division, generously made available the results of his researches into labour problems on the gold mines. Ruth Richardson in the Managing Director's office helped with secretarial services and went out of her way to remove obstacles which arose during the time of writing, as also did Jayne Moffitt, Personal Assistant to the Chairman. With gratitude I acknowledge their helpfulness, as I do also to Dorothea Jamnda, in charge of records of the Gold Division.

Advice from John Moore, the Corporation's Consulting Metallurgist, from Egon Schmidt and David Ortlepp of the Gold Division's Technical Services, undoubtedly saved me from numbers of embarrassing technical pitfalls.

Help has been no less readily forthcoming from the other side of Main Street and with appreciation I would record my indebtedness to Graham Craig and his staff in the Chairman's Office. To Betty McFarlane, Anglo's Archivist, my thanks are also due for her help.

Neville Huxham, Public Relations Manager for the Anglo American Corporation, has been closely involved with aspects of this book from its conception through to its publication. I thank him for his co-operation, made manifest in many ways.

From Denise Armstrong, Senior Librarian, and her staff in the Corporation's Library, I have received nothing but help — cheerfully given. Marlaine Botha, Photographic Librarian, has been assiduous in tracing photographs for reproduction. I would ask them all to accept my thanks for their co-operation and, no less, for the spirit in which it was rendered.

Assistance has also come from many sources outside Western Deeps and the Corporation. Notable among those who have given advice are John Imrie, Public Affairs Adviser of the South African Chamber of Mines, and his Assistant, Mrs Rosemary Burke. I am grateful also to the Chamber for permitting the use of some of its illustrative material. Tony Newby-Fraser, retired from the South African Atomic Energy Board, contributed with advice, both technical and historical. Colin Waterson, retired Managing Director of Barclays Bank, recalled attending the ceremony at which Sir Ernest inaugurated the Western Deep Levels mine. Also I thank Professor Revil Mason of the University of the Witwatersrand, and Professor Geoffrey Blainey of Melbourne University, Australia, for permitting me to quote from their published works. Lieutenant-General "Kalfie" Martin kindly granted permission to use excerpts from the South African War Histories, of which he is the Editor-in-Chief. The Standard Bank Investment Corporation generously authorised the use of extracts from a history of the Standard Bank in South Africa. With similar kindness John Lang granted me access to the manuscript of an unpublished work on Intergold, while Enid de Waal made available her unpublished thesis on the influence exercised by American engineers on developments in early deep-level operations on the Witwatersrand. "Jumbo" Pinkney gave time to recount highlights in the prominent role which South Africa played in the early development of nuclear power. Ben Ford of the South African Bureau of Standards kindly checked certain technical details for me. Viv de Chalaine, Public Relations Manager of Boart, provided useful material on modern deep drilling. The staff of the State Library in Pretoria was helpful in locating — and providing — historic material for study.

To them all I am immensely grateful. As I am no less to Professor Jack Grossert for the illustrations and decorative initials he has designed. They add very considerably to the attractiveness of the book. The manuscript benefited greatly from the experienced editing of Anthony Hamilton and for that help I am further grateful. Warm thanks also to Denise Rootenberg for her meticulous and painstaking assistance with the editing and with the reading of printer's proofs and for her co-operation, too, in the preparation of the Index. Sally Antrobus of Southern Books helped initially with the publishing of this book and similar assistance was later rendered with equal willingness and efficiency by her colleagues.

Last, but certainly not least, my warmest thanks go to Vicki, my wife, for her patience and understanding during the long months while this book was gestating: no less do I appreciate her constant encouragement and will always remain grateful for the unfailing generosity of her help in so many ways.

Bibliography

ADAMSON, R. J. *Gold Metallurgy in South Africa.* Chamber of Mines of South Africa, 1972.

ANDERSON, Ken. *Nine Flames.* Purnell, 1964.

ANTROBUS, E. S. A. *Witwatersrand Gold – 100 Years.* Geological Society of South Africa, 1968.

AXELSON, Eric. *South African Explorers.* Oxford University Press, 1953.

AXELSON, Eric. *Portuguese in South-east Africa.* Struik, Cape Town, 1973.

BARCLAYS BANK. *A Banking Centenary: 1836-1936.* Private circulation, 1936.

BATES, D. R. *The Planet Earth.* Pergamon, London, 1957.

BLAINEY, Geoffrey. *Lost causes of the Jameson Raid.* Economic History Review: Vol 18, 1965.

BREYTENBACH, J. H. *Kommandant Danie Theron.* Nasionale Boekhandel Bpk., 1950.

BREYTENBACH, J. H. *Geskiedenis van die Tweede Vryheidsoorlog.* Nasionale Boekhandel Bpk., 1949.

BUSSCHAU, W. J. *The Glamour of Gold.* SABC: Van Riebeeck Lectures, 1963.

CARTWRIGHT, A. P. *The Corner House.* Purnell, Cape Town, 1965.

CARTWRIGHT, A. P. *West Driefontein – Ordeal by Water.* Gold Fields of South Africa, 1968.

CARTWRIGHT, A. P. *Doctors of the Mines.* Purnell, 1971.

CARTWRIGHT, A. P. *Phalaborwa.* Purnell, 1972.

CHAMBERS, S. P. "Sources of Capital for Industry." *Optima:* Vol 2: No 3, 1952.

CHURCHILL, Winston. *My Early Life.* Collins, 1930.

CONNOLLY, Denis. *Guide to Southern Africa.* Connolly Publishers (Pty) Ltd, 1982.

DE KIEWIET, C. W. *A History of South Africa: Social and Economic.* Oxford University Press, 1957.

DE WAAL, Enid. *The Part Played by the Americans on the Witwatersrand During the Period 1886–1899.* Unpublished thesis for M.A.: Unisa, 1971.

DU TOIT, Alex. *The Geology of South Africa.* Oliver and Boyd, 1954.

ELLIOT, Hugh. "Down towards the Styx." *Optima:* Vol 34: No 2, 1986.

ENCYCLOPAEDIAS:
- Encyclopaedia Americana
- Encyclopaedia Britannica
- Great Soviet Encyclopedia
- New Columbia Encyclopedia
- Standard Encyclopaedia of Southern Africa

FITZPATRICK, Percy. *The Transvaal from Within.* Heinemann, London, 1899.

FRASER, Maryna (with Alan JEEVES). *All that Glittered.* O.U.P. Cape Town, 1977.

FRASER, Maryna (editor). *Johannesburg Pioneer Journals: 1888-1900.* Van Riebeeck Society, 1985.

GOLDMAN, C. S. *The Witwatersrand Gold Fields.* Effingham Wilson and Co, 1902.

GRAVES, Robert. *The Greek Myths.* Penguin, 1955.

GRAY, James. *Payable Gold.* C.N.A. Ltd, 1937.

GREEN, Timothy. *The New World of Gold.* Walker and Co, New York, 1982.

GREGORY, Theodore. *Ernest Oppenheimer.* Oxford University Press, 1962.

GUINNESS. *Book of Records.* Enfield: Guinness Superlatives, 1987.

HAGART, R. B. "The changing Pattern of gold mining Finance." *Optima:* Vol 2: No 4, 1952.

HAUGHTON, S. H. *Stratigraphic History of Africa South of the Sahara.* Oliver and Boyd, London.

HILL, F. G. "Special Techniques for Ultra-deep Mining." *Optima:* Vol 10: No 1, 1960.
HILL F. G. and MUDD J. B. *Deep Level mining Practice in South Africa.* Conference paper: Moscow, 1967.
JASTRAM, Roy W. *The Golden Constant.* John Wiley and Sons, 1977.
JASTRAM, Roy W. *The Restless Metal.* Ronald Press, 1981.
JESSUP, Edward. *Ernest Oppenheimer – a Study in Power.* Rex Collings, London, 1979.
KING, Wilfred. "The London Money Market and its Lessons for South Africa." *Optima:* Vol 6: No 4, 1956.
KNOWLES, L. G. *The VCR at Western Deep Levels. Unpublished* M.Sc. thesis: Witwatersrand University, 1966.
KRUGER, Paul. *Memoirs as told to H.C.Bredell and Piet Grobler.* Scripta Africana, 1987.
KRUGER, Rayne. *Good-bye Dolly Gray.* Cassell and Son Ltd, 1959.
LANG, John. *Bullion Johannesburg.* Jonathan Ball Publishers, Jhb, 1986.
LANG, John. *Everyman's Gold.* Unpublished ms.
MABIN, Alan (with Barbara CONRADIE). *The Confidence of the Whole Country: Reports of the Standard Bank: 1865 – 1902.* Standard Bank Investment Corporation, 1987.
MACNAB, Roy. *Gold their Touchstone.* Jonathan Ball Publishers, Jhb, 1987.
MASON, Revil. *Pre-history of the Transvaal.* Witwatersrand University Press, 1962.
McNAMARA, J.K. *Black Worker Conflicts on South African Gold Mines: 1973–1982.* Unpublished doctoral thesis: Witwatersrand University, 1985.
MEIRING, Piet. *Dynamite and Daisies.* Purnell, 1976.
MILLING-STANLEY, George. *Gold.* GFSA, 1987.
MINTER, W.E.L. "Sedimentology of the Witwatersrand Basin." *Optima:* Vol 30: No 4, 1982.
NASH, George H. *The Life of Herbert Hoover, the Engineer.* W.W. Norton and Co Inc, 1983.
NEWBY-FRASER, Tony. *Chain Reaction.* Atomic Energy Board, 1979.
OPPENHEIMER, Harry. "The Anglo American Corporation's role in South Africa's Gold mining Industry." *Optima:* Vol 34: No 2, 1986.
ORPEN, Neil (with H.J. MARTIN). *Salute the Sappers.* South African War Histories, 1984.
OSBORN, V. H. "South Africa becomes a Uranium Producer." *Optima:* Vol 3: No 1, 1953.
PLOWDEN, Edwin. "How much Uranium shall we need?" *Optima:* Vol 7: No 4, 1957.
RAYMOND, Robert. *Out of the fiery Furnace.* Macmillan Australia, 1984.
RICHARDSON, Peter (with Jean-Jacques VAN-HELTEN). *The Gold Mining Industry in the Transvaal: 1886-99.* Longman, 1980.
ROBINSON, R.E. "Ion Exchange — the Key to Uranium Production." *Optima:* Vol 6: No 4, 1956.
ROSENTHAL, Eric. *Gold! Gold! Gold!* Macmillan, 1980.
SHILLING, John W. *Historical Review of certain aspects of Taxation of Gold Mines in the Transvaal and Orange Free State.* Unpublished Dissertation: University of Witwatersrand, 1982.
SHORT, Philip. *Banda.* Routledge and Keegan Paul Ltd, 1974.
SMITH, Wilbur. *Gold Mine.* Heinemann, London, 1970.
SOUTH AFRICAN GOVERNMENT. *Official Yearbook of South Africa.* Chris van Rensburg Publications, 1986.
SUNTER, Clem. *The World and South Africa in the 1990s.* Human and Rousseau: Tafelberg, 1987.
SUTHERLAND, C. H. V. *Gold: Its Beauty, Power and Allure.* Thames and Hudson, 1959.
TARLING, D. H. *Evolution of the Earth's Crust.* Academic Press, 1978.
TROUP, Freda. *South Africa: An historical Introduction.* History Book Club, 1972.

TRUSWELL, J. F. *The geological Evolution of South Africa.* Purnell, 1977.
VAN DER POST, M. *Elandsrand.* Unpublished monograph, 1981.
VAN EEDEN, O. R. *Geology of the Republic of South Africa.* Department of Mines, 1972.
VAN RENSBURG, W. C. J. (and S. BAMBRICK). *The Economics of the World's Mineral Industries.* McGraw-Hill Book Company, 1978.
WALKER, Eric. *A History of South Africa.* Longmans, Green and Coy, 1947.
WALLIS, J. P. R (edited by Frank BRADLOW). *Thomas Baines.* A.A. Balkema, 1976.
WESTON, Rae. *Gold: a world Survey.* St Martin's Press Inc, 1983.
WILLCOX, A. R. *Southern Land: The Pre-history and History of Southern Africa.* Purnell, 1976.
WILSON, A. J. "The Pick and the Pen." *London Mining Journal,* 1979.
WILSON, Monica. *Oxford History of South Africa.* Oxford University Press, 1969.
ZEEDERBERG, Harry. *Veld Express.* Howard Timmins, Cape Town, 1971.

OFFICIAL GOVERNMENT REPORTS:
Report on the Far East Rand (UG 49 of 1925)
Commission of Enquiry into the Discovery of the Witwatersrand Goldfields: 1938.
Committee appointed by Minister of Mines; October 1943;
Chairman - Sir Robert Kotze.
Report of Committee on Deep Level Mining: January 1945.
Etc.

VARIOUS PAPERS AND PROCEEDINGS:
Chamber of Mines Reports.
Proceedings of South African Geological Society.
Reports and papers of Association of Mine Managers.
Annual Reports of Western Deep Levels.
Annual Reports of Anglo American Corporation.
Annual reports of Intergold.
Hansard; proceedings of South African Parliament.
Various newspapers, magazines, journals etc.

Index

The Genesis and Terminology of the Witwatersrand

The Witwatersrand basin and its gold fields were laid down many millions of years ago. The process of deposition was a long one. The cut away diagram depicts the scene as the last of many gold-bearing reefs — the Ventersdorp Contact Reef — was created. The various reefs were formed as rock particles eroded from northern mountains were washed down by great rivers and braided deltas. The gold-bearing particles were deposited in the river systems and their flood plane. Today these ancient river deposits, now covered by kilometres of young rock, form the rich gold-bearing reefs of the Witwatersrand.

Devised by the Geological Society of South Africa in consultation with the Chamber of Mines of South Africa with whose permission it is reproduced.